From Metal to Meat: Adapting Control Theory for the Challenges of Biological Systems

Chris

TABLE OF CONTENTS

1

Chapter 1

INTRODUCTION

Control theory originated as a tool for the design and analysis of various cyberphysical systems, including aircraft and chemical process plants. In the modern world, control theory plays a crucial role in the design and operation of robots, spacecraft, autonomous vehicles, power grids, and more. In recent years, the increased presence of *large-scale* cyberphysical systems — intelligent transport systems, multi-agent systems — has spurred renewed interest in *distributed* control theory. The first goal of this book is to present a set of theoretical advances in distributed control theory, including distributed optimal control (Chapters III and VI), distributed ro-bust control (Chapter IV), and distributed model predictive control (Chapter V). Ultimately, these theoretical advances aim to provide more scalable, efficient, and robust methods of operation for our cyberphysical systems.

The second goal of this book is to demonstrate how control theory may be applied to provide novel insights and models for biological systems. Control theory has enjoyed a few modest successes in the field of biology, but progress on this front is hindered by the incompatibility between control theory and known physiological constraints. This incompatibility arises from historical reasons: control theory was originally formulated for engineers, who typically use computers or circuit chips to implement various cyberphysical systems and controllers. Computers and circuit chips are composed of transistors and copper wires, which coordinate to produce desired signals and outputs. In organisms, cells and various chemical messengers coordinate together in a similar fashion; however, these organic components are far inferior to their metal counterparts in terms of communication speed, bandwidth, and noise. Crudely put, computing on meat is much more challenging than computing on metal, and the inability of existing control theory to capture these key biological challenges limits its applicability to biology.[1] Thus, the second goal of this book is to present a set of theoretical advances (including distributed control theory) that begin to bridge this gap between control theory and biology, and to demonstrate how these theoretical advances can provide insights on sensorimotor structures in the primate cortex (Chapter VI) and models of fruit fly locomotion (Chapter VII).

---

[1] This problem is shared by the more popular neighbor of control theory: machine learning (more specifically, neural networks).

At this point, the reader will likely not require additional persuasion regarding the first goal of this book; its motivations are standard and well-aligned with the goals of control theory and engineering as a whole. However, the second goal — a somewhat unconventional union between control theory and biology — may raise the eyebrows of both engineers and biologists alike. I will now elaborate on this goal for engineers who may find this choice of application puzzling (in other words, "why biology?"), and also for biologists who may be skeptical as to how this theory can be of use to them (in other words, "why control theory?").

1.1 Why Biology?

The pursuit of autonomy and artificial intelligence — particularly in the domain of robotics — is a major driving force behind funding, research, and technological advances in both academia and industry. However, despite significant investments and breakthroughs, state-of-the-art systems are often bested by biology. One only has to look across the street at the neighbor's cat to see an example of an autonomous being that is at once more agile, energy-efficient, and robust to environmental variations than our best quadruped robots. In general, animals excel at complex tasks (e.g. predator evasion) in highly dynamic and unpredictable environments — modern engineering systems, for the most part, cannot achieve comparable performance for even simple subsets of these tasks (e.g. legged locomotion, balancing). This contrast is made more remarkable by the fact that, as previously mentioned, animals are composed of organic components that are in many ways inferior to the metal components that make up our robots. In short, animals attain a level of robustness and autonomy that engineers have thus far failed to replicate. By gaining a deeper understanding of how biology accomplishes this, engineers can gain valuable insights to aid in their designs.

The study of sensorimotor systems — that is, how organisms use sensory information (e.g. vision, hearing, proprioception) to inform decisions and movements — is highly relevant to the presented motivation. Fortuitously, this is the subfield of biology in which control theory has found the strongest foothold; indeed, the results in this book focus on various aspects of the sensorimotor system. The general approach is to utilize control theory to build new models or supplement existing models in order to gain insights into the design principles of biology. Given some dynamical system, objectives, and constraints, we can use control theory to provide mathematical descriptions of optimal system operation and behavior. This can model the behavior of organisms, as well as the physiology — if we are clever about the formulation. In addition to sensorimotor systems, many other biological processes (e.g. biomolecular chemical reactions, biocircuits) can be modeled using the tools of ordinary differential equations and dynamical systems. Control theory is a natural extension of dynamical systems theory, and may be applicable to these biological processes as well; this is the subject of ongoing work.

In the realm of engineering, we typically have well-established systems, objectives, and constraints — the main research problem is how to design the controller to satisfy requirements. In the realm of biology, things are not quite so clear cut; typically, neither system, objectives, nor constraints will be obvious to us. Organisms are complex systems, and we typically only have sparse observations on these systems. For instance, even a small fruit fly has hundreds of thousands of neurons, which coordinate in various ways to produce a rich repertoire of behaviors. However, data on this organism is typically limited to specific clusters of neurons or specific behavioral contexts. This observational sparsity is challenging to engineering researchers unfamiliar with biology, and requires a change in mindset: instead of demanding details that are experimentally costly or impossible to obtain, we want to think about what can be done with the data that is available. This requires making simplifying abstractions, which allow us to produce methods and insights that are dependent on fundamental system properties rather than unknown details. For example, in Chapter VI, the physiological insights arise from the inclusion of sparse communication constraints, and not on any specific numerical quantities. Of course, we can also mix data-driven techniques and control theory to produce novel quantitative models, as is done in Chapter VII.

In general, the formulation of control theory for biology problems varies greatly from project to project, and is also quite different from the formulation of control theory for engineering problems. Even even within the field of biology, different problems may require different approaches. However, it is not the approach or formulation that matters, but the biological research question — and it is crucial that this question is formulated with input from actual biologists. It is no mean feat to find a problem that is both biologically meaningful and appropriate for control theory, but when we do, the results are uniquely rewarding. I must also caution that it is thoroughly unrewarding to work on biology-inspired problems that lack these qualities. Laboring away at a contrived bio-inspired problem that has neither applicability to biology nor engineering is quite meaningless; attempting to shoehorn control theory into a biology problem for which a different mathematical tool would be better suited is tedious and unfruitful. Good collaborators in biology will help one to avoid the former, and it is an interdisciplinary exercise to avoid the latter.

Another reason why biology should be of interest to control theorists is the relative maturity of control theory as a field. Theoretical progress is often driven by applications, and applications of control theory in industry typically use techniques developed decades ago. Exploring new application directions in biology can raise new theoretical questions with both intellectual and practical merit — for instance, how to implement a controller with signaling limitations.

In summary, applying control theory to biology has two main benefits. Firstly, a better understanding of biological autonomy can provide design insights for engineers who wish to construct autonomous systems. Secondly, a fresh application area introduces new and interesting theoretical problems.

1.2 Why Control Theory?

I start with an obvious caveat: control theory is not applicable to the entirety of biology. However, it is a cornerstone of models in the area of sensorimotor neuroscience, as mentioned above. Adjacent fields, such as motor control and locomotion, naturally also benefit from control theory. For biologists working in more distant areas, a general guideline is that if the biological entities of interest can be described with differential equations, then applying concepts from dynamical systems and control theory may provide additional insights and models.

Control theory is fundamentally concerned with *optimization*: given some dynamical system, objective, and constraints, what is the best behavior possible, and how is this behavior obtained? This is a natural way of looking at biological systems, which are optimized for various survival-critical tasks through evolution. Interpreting biology from this perspective offers two key benefits: prior modeling and human-interpretable models. Firstly, control-based models can be formulated as *prior* models — that is, they can predict experimental outcomes before the experiment is conducted. This is complementary to data-driven models, which provide insights via analysis only after experiments have been performed. Prior models can identify potentially interesting future experiments; this is particularly important as the space of available experiments grows exponentially with the advancement of experimental techniques. Additionally, control theory creates human-interpretable models from human-interpretable inputs (i.e. system, objective, and constraints), which can reveal fundamental principles that govern complex systems — an example is given in Chapter VII.

In the domain of sensorimotor neuroscience, control theory also provides the mathematical tools needed to translate longstanding qualitative ideas on closed-loop sensing and action into quantitative models. These ideas are concerned with the analysis of sensory areas (e.g. vision) not as standalone modules, but as part of the larger sensorimotor feedback loop. The general underlying principle is that all neural processes serve motor output; sensory processing is only useful insofar as it produces appropriate actions. While analysis of solitary modules such as vision are dominated (and rightfully so) by other mathematical approaches, the analysis of the full loop necessitates the use of control theory, which is fundamentally concerned with the closed-loop relationship between sensing, action, and environment. Unique insights can come of looking at the brain from this closed-loop perspective as opposed to individual modules, as shown in Chapter VI.

Chapter 2

TECHNICAL SETUP

2.1 Notation and Abbreviations

- Italicized lower-case letters (e.g. x) denote scalars or vectors in the time domain. They may also denote functions — the distinction will be apparent from context.

- Italicized upper-case letters (e.g. A) denote constant matrices.

- Calligraphic upper-case letters (e.g. $\mathcal{P}$) denote sets.

- Fraktur upper-case letters (e.g. $\mathfrak{R}$) denote some subset of $\mathbb{Z}^+$, e.g. $\mathfrak{R} = \{1, \ldots, n\}, n \in \mathbb{Z}^+$.

- An arrow above a matrix quantity denotes vectorization, i.e. $\vec{A}$ is the vectorization of A.

- Round brackets with subscripts denote elements of matrices and vectors. For example, $(A)_{i,j}$ denotes the element in the ith row, j^{th} column of A. $(A)_{i,:}$ denotes the ith row of A. $(A)_{i:,:}$ denotes the rows of A starting from the ith row. Subscripts can also be sets: $(A)_{\mathfrak{R},\mathfrak{C}}$ denotes the submatrix of A composed of the rows and columns specified by $\mathfrak{R}$ and $\mathfrak{C}$, respectively. Where appropriate, we may also overload subscript notation to indicate block partitions of a matrix; the meaning will be clear from context. Occasionally for vectors, we will omit the round brackets for notational simplicity.

- Subscripts without round brackets have several uses, all of which should be apparent from context. They are most commonly used to denote block partitions of matrices or vectors. Additionally, in Chapter V, we will use subscripts x_t to denote the *predicted* value x in t timesteps (not to be confused with $x(\tau)$, which is the true value of x at time τ.

- Square brackets with subscripts denote elements of matrices and vectors corresponding to specific subsystems. For example, $[A]_{i,j}$ represents influence from the jth subsystem to the ith subsystem. This notation is convenient when subsystems may contain multiple states or inputs.

- Boldface lower-case letters (e.g. $\mathbf{x}$) denote signals in the discrete-time frequency domain, i.e. z-domain.

- Boldface upper-case letters (e.g. $\boldsymbol{\Phi}$) denotes transfer matrices in the frequency domain, i.e. z-domain. Transfer matrices are composed of spectral elements (constant matrices). For example, transfer matrix $\boldsymbol{\Phi}$ may be written as $\boldsymbol{\Phi} := \sum_{k=0}^{T} \Phi(k) z^{-k}$, where $\Phi(k)$ is the kth spectral element of $\boldsymbol{\Phi}$, and $T \in \mathbb{Z}_+$ is the finite-impulse-response (FIR) horizon length. Subscript-style notation also applies to boldface transfer matrices. For example, $[\boldsymbol{\Phi}]_{i,j} := \sum_{k=0}^{T} [\Phi(k)]_{i,j} z^{-k}$.

- Superscript ^{blk} indicates augmented matrices. For any matrix Z, the corresponding augmented matrix Z^{blk} is defined as a block-diagonal matrix containing N_x copies of Z, i.e. $Z^{\text{blk}} := \text{blkdiag}(Z, \ldots Z)$. For any matrix $Y = Z\Lambda$, the corresponding vectorization can be written as $\vec{Y} = Z^{\text{blk}} \vec{\Lambda}$.

For the most part, we will omit dimensions; when omitted, they will be obvious from context and assumed to be compatible.

Table 2.1: Common abbreviations

FIR	Finite impulse response
LQR	Linear quadratic regulator
LQG	Linear quadratic Gaussian
MPC	Model predictive control
SLS	System level synthesis

2.2 System Level Synthesis

Chapters III, IV, V, and VI of this book make use of the system level synthesis (SLS) framework. In this section, we introduce the basics of SLS, adapting material from [1]–[3]. The interested reader is referred to the review paper [3] for more details.

Consider the linear time-invariant dynamics

$$x(t + 1) = Ax(t) + Bu(t) + w(t) \tag{2.1}$$

where x, u, and w denote state, control input, and exogenous disturbance, respectively.

We can rewrite this in z-domain by applying the z-transform, the discrete-time analogue of the Laplace transform, to obtain

$$z\mathbf{x} = A\mathbf{x} + B\mathbf{u} + \mathbf{w} \tag{2.2}$$

and apply linear causal state feedback controller $\mathbf{u} = \mathbf{Kx}$. Define $\boldsymbol{\Phi}_x$ and $\boldsymbol{\Phi}_u$, transfer matrices representing closed-loop responses (also referred to as closed-loop maps) from disturbance to state and control

$$\begin{bmatrix} \mathbf{x} \\ \mathbf{u} \end{bmatrix} = \begin{bmatrix} (zI - A - B\mathbf{K})^{-1} \\ \mathbf{K}(zI - A - B\mathbf{K})^{-1} \end{bmatrix} \mathbf{w} =: \begin{bmatrix} \boldsymbol{\Phi}_x \\ \boldsymbol{\Phi}_u \end{bmatrix} \mathbf{w} \tag{2.3}$$

Theorem 2.1. (Theorem 4 in [3]) For system (2.2) with linear state-feedback controller $\mathbf{u} = \mathbf{Kx}$, the following hold:

1. The affine subspace

$$\begin{bmatrix} zI - A & -B \end{bmatrix} \begin{bmatrix} \boldsymbol{\Phi}_x \\ \boldsymbol{\Phi}_u \end{bmatrix} = I, \quad \boldsymbol{\Phi}_x, \boldsymbol{\Phi}_u \in \frac{1}{z}\mathcal{RH}_\infty \tag{2.4}$$

   parametrizes all responses $\boldsymbol{\Phi}_x, \boldsymbol{\Phi}_u$ achievable by an internally stabilizing state feedback controller $\mathbf{K}$. This is the *state feedback achievability constraint*. Here, $\frac{1}{z}\mathcal{RH}_\infty$ is the set of stable, strictly causal LTI transfer matrices.

2. For any $\boldsymbol{\Phi}_x, \boldsymbol{\Phi}_u$ obeying (2.4), the controller $\mathbf{K} = \boldsymbol{\Phi}_u \boldsymbol{\Phi}_x^{-1}$, implemented as

$$\hat{\boldsymbol{\delta}} = \mathbf{x} + (I - z\boldsymbol{\Phi}_x)\hat{\boldsymbol{\delta}}, \quad \mathbf{u} = z\boldsymbol{\Phi}_u\hat{\boldsymbol{\delta}} \tag{2.5}$$

   achieves the desired closed-loop response as per (2.3).

Theorem 2.1 allows us to parametrize control problems as optimizations over closed-loop responses $\boldsymbol{\Phi}_x$ and $\boldsymbol{\Phi}_u$, instead of controller $\mathbf{K}$. For example, we can write the LQR problem with state penalty matrix Q and input penalty matrix R using SLS as

$$\min_{\boldsymbol{\Phi}_x, \boldsymbol{\Phi}_u} \left\| \begin{bmatrix} Q^{\frac{1}{2}} & 0 \\ 0 & R^{\frac{1}{2}} \end{bmatrix} \begin{bmatrix} \boldsymbol{\Phi}_x \\ \boldsymbol{\Phi}_u \end{bmatrix} \right\|_F^2 \tag{2.6a}$$

$$\text{s.t.} \quad \begin{bmatrix} zI - A & -B \end{bmatrix} \begin{bmatrix} \boldsymbol{\Phi}_x \\ \boldsymbol{\Phi}_u \end{bmatrix} = I \tag{2.6b}$$

$$\boldsymbol{\Phi}_x, \boldsymbol{\Phi}_u \in \frac{1}{z}\mathcal{RH}_\infty \tag{2.6c}$$

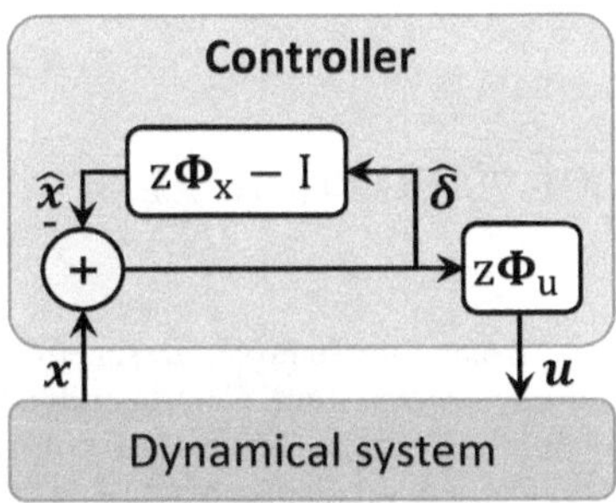

Figure 2.1: Implementation of a state feedback system level synthesis controller. Φ_x and Φ_u are the closed-loop responses from exogeneous disturbance to state and input, respectively.

where $\| \cdot \|_F$ indicates the Frobenius norm. Example of how to formulate other standard control problems (e.g. LQG) as optimizations over closed-loop responses are given in [3]. Once the optimal closed-loop responses are found, they can be used to implement the controller via (2.5). This is pictorially depicted in Figure 2.2.

The advantages of the the SLS parametrization are twofold. Firstly, this parametrization allows us to incorporate delayed communication, local communication (in which distributed agents are allowed only to communicate to other agents within some local neighborhood), and localized disturbance rejection (in which the effects of a disturbance at a given agent is forced to be contained within some local neighborhood of the agent) into the controller synthesis problem. In the SLS formulation, these constraints manifest themselves as sparsity constraints on decision variables Φ_x and Φ_u. These sparsity constraints are affine and easy to enforce, allowing the resulting problem to be solved using tools from convex optimization. Secondly, this formulation allows us to perform both controller synthesis *and* controller implementation in a distributed manner, conferring considerable scalability benefits. For controller synthesis, we can decompose the full optimization problem into sub-problems to be solved in parallel by each subsystem, resulting in a synthesis procedure with $O(1)$ complexity relative to network size — more details are described in Chapter IV. In the remainder of this book, we leverage these advantages for both theoretical advances in controls and conceptual insight for biological systems.

Chapter 3

CLOSED-LOOP VS. CONTROLLER SPECIFICATIONS

[1]   J. S. Li and D. Ho, "Separating Controller Design from Closed-Loop Design: A New Perspective on System-Level Controller Synthesis," in *IEEE American Control Conference*, 2020, pp. 3529–3534. DOI: 10.23919/ACC45564. 2020.9147736. [Online]. Available: http://arxiv.org/abs/2006. 05040,

Overview: We show that given a desired closed-loop response for a system, there exists an affine subspace of controllers that achieve this response. By leveraging the existence of this subspace, we are able to separate controller design from closed-loop design by first synthesizing the desired closed-loop response and then synthesizing a controller that achieves the desired response. This is a useful extension to the SLS framework, in which the controller and closed-loop response are jointly synthesized and we cannot enforce controller-specific constraints without subjecting the closed-loop response to the same constraints. We demonstrate the importance of separating controller design from closed-loop design with an example in which communication delay and locality constraints cause standard SLS to be infeasible. Using our new two-step procedure, we are able to synthesize a controller that obeys communication constraints while only incurring a 3% increase in LQR cost compared to the optimal LQR controller. Overall, our proposed two-step synthesis allows us to design low-cost, distributed controllers that were unavailable to us in the previous framework.

3.1   Introduction

In the standard SLS framework, the closed-loop responses Φ_x and Φ_u are directly used to implement the controller. Thus, any constraints applied to the controller are enforced on the closed-loop response as well. For example, we cannot enforce local communication constraints on the controller without also enforcing localized disturbance rejection on the closed-loop response. This can be overly restrictive, especially in situations where we are only concerned with communication constraints and not with closed-loop constraints. In the worst case, standard SLS can be infeasible under strict communication constraints, because SLS will also by definition enforce analogous closed-loop constraints, and the latter is often excessively strict. This is partially addressed by virtually local SLS [4], which searches over approxi-

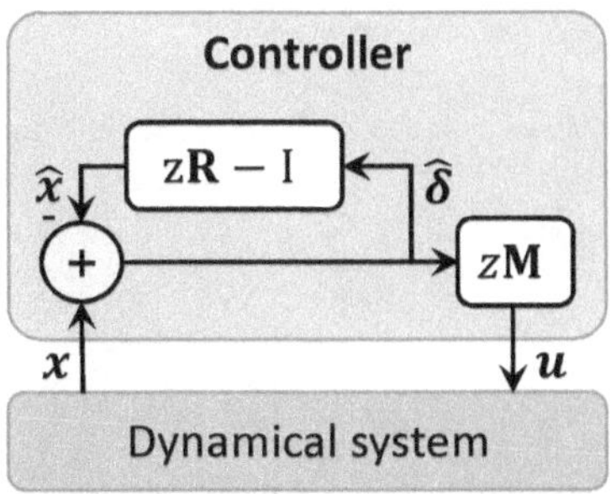

Figure 3.1: Implementation of state feedback controller.

mate closed-loop responses instead of exact closed-loop responses; constraints are imposed on the approximate closed-loop responses, which mitigates infeasibility. In this chapter, we propose an alternative two-step procedure:

1. Synthesize the desired closed-loop response, subject to closed-loop constraints. This can be done using SLS or any other linear synthesis method.

2. Synthesize the controller, subject to controller constraints.

This method confers performance benefits over the virtually local method, as demonstrated in the simulations section. Additionally, this method allows us to cleanly separate controller specifications from closed-loop specifications, an important semantic difference for many applications.

3.2 Implementation Matrices

To fully separate closed-loop response constraints from controller constraints, we require a controller that is implemented using transfer matrices *other* than the closed-loop responses. Figure 3.1 shows the controller implementation. $\mathbf{R}$ and $\mathbf{M}$ are the controller implementation matrices, with horizon T_c. In standard SLS, we set $\mathbf{R} = \mathbf{\Phi}_x$ and $\mathbf{M} = \mathbf{\Phi}_u$. In this chapter, we will explore the alternatives.

The controller contains two internal signals: $\mathbf{x}$ and $\hat{\delta}$. The frequency-domain equations describing the controller are

$$\hat{\delta} = \mathbf{x} + (I - z\mathbf{R})\hat{\delta} \tag{3.1a}$$

$$\mathbf{x} = z\mathbf{R}\hat{\delta} \tag{3.1b}$$

$$\mathbf{u} = z\mathbf{M}\hat{\boldsymbol{\delta}} \qquad (3.1c)$$

Assuming that $R(1) = I$, we can rewrite these in the time domain:

$$\hat{\delta}(t) = x(t) - \sum_{k=2}^{T_c} R(k)\hat{\delta}(t - k + 1) \qquad (3.2a)$$

$$u(t) = \sum_{k=1}^{T_c} M(k)\hat{\delta}(t - k + 1) \qquad (3.2b)$$

we remark that this assumption will not limit the space of available $\mathbf{R}$ and $\mathbf{M}$; equivalent results can be obtained with arbitrary $R(1)$, although the bookkeeping of variables becomes more tedious in this case.

Proposition 3.1. Any linear law $\mathbf{u} = \mathbf{Kx}$ can be implemented using the controller structure defined in Figure 3.1.

Proof. Construct closed $\boldsymbol{\Phi}_x$ and $\boldsymbol{\Phi}_u$ directly from $\mathbf{K}$ via (2.3):

$$\begin{bmatrix} \boldsymbol{\Phi}_x \\ \boldsymbol{\Phi}_u \end{bmatrix} = \begin{bmatrix} (zI - A - BK)^{-1} \\ \mathbf{K}(zI - A - BK)^{-1} \end{bmatrix} \qquad (3.3)$$

Set $\mathbf{R} = \boldsymbol{\Phi}_x$ and $\mathbf{M} = \boldsymbol{\Phi}_u$ in (3.1). This recovers the control law $\mathbf{u} = \mathbf{Kx}$. $\quad\square$

Controllers and closed-loop responses

Lemma 3.1. Let $(\boldsymbol{\Phi}_x, \boldsymbol{\Phi}_u)$ be stable closed-loop responses. The only linear controller $\mathbf{K}$ (i.e. $\mathbf{u} = \mathbf{Kx}$) that achieves these closed-loop responses is $\mathbf{K} = \boldsymbol{\Phi}_u\boldsymbol{\Phi}_x^{-1}$.

Proof. By Theorem 2.1, $\mathbf{K} = \boldsymbol{\Phi}_u\boldsymbol{\Phi}_x^{-1}$ achieves the closed-loop responses. We show uniqueness by contradiction. Assume there is another linear controller $\mathbf{K}_1 \neq \mathbf{K}$ that also achieves the desired closed-loop responses. Since both $\mathbf{K}$ and $\mathbf{K}_1$ achieve $(\boldsymbol{\Phi}_x, \boldsymbol{\Phi}_u)$, we have that

$$\boldsymbol{\Phi}_x = (zI - A - BK_1)^{-1} = (zI - A - BK)^{-1} \qquad (3.4a)$$

$$\boldsymbol{\Phi}_u = \mathbf{K}_1(zI - A - BK_1)^{-1} = \mathbf{K}(zI - A - BK)^{-1} \qquad (3.4b)$$

Substituting (3.4a) into (3.4b) gives

$$\mathbf{K}_1\boldsymbol{\Phi}_x = \mathbf{K}\boldsymbol{\Phi}_x \qquad (3.5)$$

Since $\boldsymbol{\Phi}_x$ is invertible, this implies that $\mathbf{K}_1 = \mathbf{K}$. Contradiction! $\quad\square$

From Lemma 3.1 and the definitions of the closed-loop responses (2.3), we see that there is a one-to-one mapping between a given linear controller and its closed-loop responses. However, the linear controller $\mathbf{K}$ can be *implemented* in a variety of ways. For example, we could directly implement $\mathbf{u} = \mathbf{K}\mathbf{x}$; we could also implement a linear controller using the structure shown in Figure 3.1. In the original SLS framework, the latter is used to avoid direct matrix inversion of $\mathbf{\Phi}_x$.

Implementing closed-loop responses

We introduce some necessary terminology:

Definition 3.1. For the controller structure shown in Figure 3.1 with controller matrices $\mathbf{R}$ and $\mathbf{M}$, let the resulting closed-loop response be $(\tilde{\mathbf{\Phi}}_x, \tilde{\mathbf{\Phi}}_u)$. Then, we say that $(\mathbf{R}, \mathbf{M})$ are *implementation matrices* for $(\tilde{\mathbf{\Phi}}_x, \tilde{\mathbf{\Phi}}_u)$, and $(\tilde{\mathbf{\Phi}}_x, \tilde{\mathbf{\Phi}}_u)$ are the *implemented closed-loop responses* of controller matrices $\mathbf{R}$ and $\mathbf{M}$.

The implemented closed-loop responses are found by combining (3.2) and (2.1):

$$\begin{bmatrix} \tilde{\mathbf{\Phi}}_x \\ \tilde{\mathbf{\Phi}}_u \end{bmatrix} = \begin{bmatrix} \mathbf{R} \\ \mathbf{M} \end{bmatrix} \mathbf{\Delta}_c^{-1}, \quad \mathbf{\Delta}_c = \begin{bmatrix} zI - A & -B \end{bmatrix} \begin{bmatrix} \mathbf{R} \\ \mathbf{M} \end{bmatrix} \tag{3.6}$$

Note that $\mathbf{\Delta}_c$ can also be written as $I + \mathbf{\Delta}$. This is the same formulation used by (4.22) in [3], modulo notational differences (we use $\mathbf{R}$ and $\mathbf{M}$ instead of $\hat{\mathbf{\Phi}}_x$, $\hat{\mathbf{\Phi}}_u$). $\mathbf{\Delta}_c$ is invertible since its leading spectral element, I, is invertible.

Our analysis largely focuses on closed-loop responses $(\mathbf{\Phi}_x, \mathbf{\Phi}_u)$ instead of the controller $\mathbf{K}$. We remark that we can also view $(\mathbf{R}, \mathbf{M})$ as implementation matrices for the controller $\mathbf{K} = \mathbf{\Phi}_u \mathbf{\Phi}_x^{-1}$, since there is a one-to-one mapping between controller and closed-loop responses.

Theorem 3.1. For $R(1) = I$, $(\mathbf{R}, \mathbf{M})$ are implementation matrices for $(\mathbf{\Phi}_x, \mathbf{\Phi}_u)$ *if and only if* they satisfy the following *implementation constraint*

$$\begin{bmatrix} \mathbf{R} \\ \mathbf{M} \end{bmatrix} = \begin{bmatrix} \mathbf{\Phi}_x \\ \mathbf{\Phi}_u \end{bmatrix} \begin{bmatrix} zI - A & -B \end{bmatrix} \begin{bmatrix} \mathbf{R} \\ \mathbf{M} \end{bmatrix} \tag{3.7}$$

Proof. Necessity. If $(\mathbf{R}, \mathbf{M})$ are implementation matrices for $(\mathbf{\Phi}_x, \mathbf{\Phi}_u)$, then

$$\begin{bmatrix} \tilde{\mathbf{\Phi}}_x \\ \tilde{\mathbf{\Phi}}_u \end{bmatrix} = \begin{bmatrix} \mathbf{R} \\ \mathbf{M} \end{bmatrix} \left(\begin{bmatrix} zI - A & -B \end{bmatrix} \begin{bmatrix} \mathbf{R} \\ \mathbf{M} \end{bmatrix} \right)^{-1} = \begin{bmatrix} \mathbf{\Phi}_x \\ \mathbf{\Phi}_u \end{bmatrix} \tag{3.8}$$

where the first equality arises from (3.6), and the second equality arises from the fact that the implemented closed-loop responses must be equal to (Φ_x, Φ_u). Multiplying by $\begin{bmatrix} zI - A & -B \end{bmatrix} \begin{bmatrix} \mathbf{R} \\ \mathbf{M} \end{bmatrix}$ gives the desired result.

Sufficiency. If $(\mathbf{R}, \mathbf{M})$ satisfy (3.7), we can substitute (3.7) into (3.6) to conclude that $(\tilde{\Phi}_x, \tilde{\Phi}_u) = (\Phi_x, \Phi_u)$, i.e. $(\mathbf{R}, \mathbf{M})$ are implementation matrices for (Φ_x, Φ_u), as desired. $\qquad\square$

Theorem 3.1 describes an affine subspace of implementation matrices for (Φ_x, Φ_u).

Corollary 3.1. If $(\mathbf{R}, \mathbf{M})$ are implementation matrices for (Φ_x, Φ_u), then the first spectral components of $\mathbf{M}$ and Φ_u are equal, i.e. $M(1) = \Phi_u(1)$.

Proof. If $(\mathbf{R}, \mathbf{M})$ are implementation matrices for (Φ_x, Φ_u), then (3.7) must hold. Writing out (3.7) in terms of its spectral elements gives the desired result. $\qquad\square$

Corollary 3.2. For $T_c \geq T$, (Φ_x, Φ_u) are implementation matrices for themselves.

This corollary agrees with results from standard SLS, in which (Φ_x, Φ_u) are used as implementation matrices as well as closed-loop responses.

Corollary 3.3. If $(\mathbf{R}, \mathbf{M})$ are implementation matrices for (Φ_x, Φ_u), then controller $\mathbf{K} = \Phi_u \Phi_x^{-1} = \mathbf{M}\mathbf{R}^{-1}$

Proof. The first equality holds by Theorem 2.1. To see that the second equality holds, combine equations (3.1b) and (3.1c). $\qquad\square$

Existence of solutions

Corollary 3.2 states that (3.7) has at least one solution for $T_c \geq T$. We now provide some analysis on the existence of solutions when $T_c < T$. In this case, we cannot directly work with the transfer matrices, as they have unequal horizon lengths — we will instead write out block matrices of spectral elements. We first define some block matrices:

$$\overline{R} = \begin{bmatrix} R(1) \\ R(2) \\ \vdots \\ R(T_c) \end{bmatrix}, \overline{M} = \begin{bmatrix} M(1) \\ M(2) \\ \vdots \\ M(T_c) \end{bmatrix}, Z_{AB} = \begin{bmatrix} I & & & 0 & \\ -A & \ddots & & -B & \\ & \ddots & I & & \ddots \\ & & -A & & -B \end{bmatrix} \tag{3.9a}$$

$$
\overline{\Phi_x} = \begin{bmatrix} \Phi_x(1) & & & & \\ \Phi_x(2) & \ddots & & & \\ \vdots & & \ddots & & \\ \Phi_x(T) & & & & \\ & & & \ddots & \\ & & & & \Phi_x(T) \end{bmatrix}, \overline{\Phi_u} = \begin{bmatrix} \Phi_u(1) & & & & \\ \Phi_u(2) & \ddots & & & \\ \vdots & & \ddots & & \\ \Phi_u(T) & & & & \\ & & & \ddots & \\ & & & & \Phi_u(T) \end{bmatrix} \tag{3.9b}
$$

Then, (3.7) can be rewritten in block matrix form as

$$
\begin{bmatrix} \overline{R} \\ 0 \\ \overline{M} \\ 0 \end{bmatrix} = \begin{bmatrix} \overline{\Phi_x} \\ \overline{\Phi_u} \end{bmatrix} Z_{AB} \begin{bmatrix} \overline{R} \\ \overline{M} \end{bmatrix} \tag{3.10}
$$

where the zeros on the left hand side arise from the fact that $T_c < T$. We can split constraint (3.10) into two parts. The first part can be written as:

$$
\begin{bmatrix} \overline{R} \\ \overline{M} \end{bmatrix} = \begin{bmatrix} (\overline{\Phi_x})_1 \\ (\overline{\Phi_u})_1 \end{bmatrix} Z_{AB} \begin{bmatrix} \overline{R} \\ \overline{M} \end{bmatrix} \tag{3.11}
$$

where we have partitioned $\overline{\Phi_x} = \begin{bmatrix} (\overline{\Phi_x})_1 \\ (\overline{\Phi_x})_2 \end{bmatrix}$ with appropriate dimensions, i.e. $(\overline{\Phi_x})_1$ has the same number of rows as $\overline{R}$. A similar partition is applied to $\overline{\Phi_u} = \begin{bmatrix} (\overline{\Phi_u})_1 \\ (\overline{\Phi_u})_2 \end{bmatrix}$, where $(\overline{\Phi_u})_1$ has the same number of rows as $\overline{M}$. Constraint (3.11) can be further rearranged into

$$
\left(\begin{bmatrix} (\overline{\Phi_x})_1 \\ (\overline{\Phi_u})_1 \end{bmatrix} Z_{AB} - I \right) \begin{bmatrix} \overline{R} \\ \overline{M} \end{bmatrix} = 0 \tag{3.12}
$$

The second part of (3.10) can be written as

$$
\begin{bmatrix} (\overline{\Phi_x})_2 \\ (\overline{\Phi_u})_2 \end{bmatrix} Z_{AB} \begin{bmatrix} \overline{R} \\ \overline{M} \end{bmatrix} = 0 \tag{3.13}
$$

Stacking (3.12) and (3.13) together, we can write

$$F \begin{bmatrix} \overline{R} \\ \overline{M} \end{bmatrix} = 0, \quad F = \begin{bmatrix} \begin{bmatrix} (\overline{\Phi_x})_1 \\ (\overline{\Phi_u})_1 \end{bmatrix} Z_{AB} - I \\ \begin{bmatrix} (\overline{\Phi_x})_2 \\ (\overline{\Phi_u})_2 \end{bmatrix} Z_{AB} \end{bmatrix} \tag{3.14}$$

Recall that the first spectral element $R(1)$ is restricted to be equal to the identity. Then, we can partition $F = \begin{bmatrix} F_1 & F_2 \end{bmatrix}$ where the number of columns in F_1 is equal to the number of rows in $R(1)$. Finally, we introduce free variable $v := \begin{bmatrix} R(2) \\ \vdots \\ R(T_c) \\ \overline{M} \end{bmatrix}$ and rewrite (3.7) as $F_2 v = -F_1$, where F_1 and F_2 are constant for some given system and closed loop maps, and v contains the variables representing the implementation matrices. This is a simple affine constraint. To check whether solutions (i.e. implementation matrices) exist, we need only check the conditions of the Rouché-Capelli theorem [5] — namely, that $\text{rank}(F_2) = \text{rank}(F)$.

3.3 Stability

Internal dynamics

The system is internally stable if the dynamics of $\hat{\delta}$, the internal signal, are stable. Substitute (3.2) into (2.1) to obtain internal dynamics of the form

$$z_t = \begin{bmatrix} \hat{\delta}_{t-T_c+1} \\ \vdots \\ \hat{\delta}_{t-1} \\ \hat{\delta}_t \end{bmatrix}, \quad z_{t+1} = A_z z_t, \quad A_z = \begin{bmatrix} 0 & I & \dots & 0 & 0 \\ \vdots & & & & \ddots \\ 0 & 0 & \dots & 0 & I \\ -\Delta_c(T_c) & & \dots & & -\Delta_c(1) \end{bmatrix} \tag{3.15}$$

where the spectral elements of $\mathbf{\Delta}_c$ depend on the implementation matrices $(\mathbf{R}, \mathbf{M})$.

Stability check

We can verify internal stability *a posteriori* by checking that A_z is stable. Alternatively, a sufficient condition for internal stability is $\|\mathbf{\Delta} = \mathbf{\Delta}_c - I\| < 1$ [4].

The stability of A_z can be checked in a distributed manner. First, a helpful proposition:

Proposition 3.2. Let $\| \cdot \|$ be an induced matrix norm. For $A \in \mathbb{R}^{n \times n}$, if $\exists m > 0$ s.t. $\|A^m\| < 1$, then A is stable.

Proof. Let $\rho = \|A^m\|^{1/m}$, $\rho \in [0, 1)$. Using norm submultiplicativity, we can show that $\forall t > m$, $\|A^t\| \leq C\rho^t$ for some constant C. Using this upper bound and induced norm properties, we can show that $\forall x_0 \in \mathbb{R}^n$, $\lim_{t \to \infty} \|A^t x_0\| = 0$. This is the definition of stability in the discrete-time setting. $\square$ $\qquad\qquad\qquad\square$

Due to the structure of $\mathbf{\Delta}_c$, the internal dynamics matrix A_z can be computed in parallel using local information only. After A_z has been computed, we let each subsystem i store A_z its its entirety. The distributed stability check procedure is as follows:

Algorithm 3.1 Distributed stability check for alternative implementation matrices

---

    input : A_z, k_{max}, n_{max}
    output : Boolean value indicating whether internal stability was certified
    for $k = 1 \dots k_{max}$ **:**
1:    For every subsystem i, $[A_z^k]_i \leftarrow A_z [A_z^{k-1}]_i$
2:    **if** $\|[A_z^k]_i\|_{1 \to 1} < 1$ for all i **:**
        return True
    elseif $\exists$ i s.t. $\|[A_z^k]_i\|_{1 \to 1} > n_{max}$ **:**
        return False

Here, k_{max} is some predetermined maximum number of iterations, and n_{max} is some predetermined threshold for the transient condition; since $\|A_z^k\|$ corresponds to the amplitude of the transient response, this termination condition involving n_{max} corresponds to finding an unacceptably large transient condition.

We use the induced 1-to-1 norm for its column-wise separability; all computations can be distributedly computed, though the final stability certification requires consensus. In simulations later in this chapter, this algorithm certifies stability in 7 iterations for the low-order controller and 32 iterations for the full-order controller.

3.4 Approximate Implementations

In this section, we introduce relaxations of (3.7). These relaxations are preferable for two reasons. Firstly, the original constraint (3.7) may not permit sparse implementation matrices for certain closed-loop responses. For instance, if $\Phi_u(1)$ is dense, then $M(1)$ must also be dense, according to Corollary 3.1. Secondly, we find that enforcing (3.7) often yields implementation matrices that are internally unstable.

Consider the implemented closed-loop responses $(\tilde{\Phi}_x, \tilde{\Phi}_u)$ of implementation matrices $(\mathbf{R}, \mathbf{M})$. Instead of enforcing $(\tilde{\Phi}_x, \tilde{\Phi}_u)$ to be equal to (Φ_x, Φ_u) via (3.7), we now only want them to be close under some norm, i.e. we penalize $\left\| \begin{bmatrix} \Phi_x \\ \Phi_u \end{bmatrix} - \begin{bmatrix} \tilde{\Phi}_x \\ \tilde{\Phi}_u \end{bmatrix} \right\|$, which can be rewritten using (3.6) as

$$\left\| \begin{bmatrix} \Phi_x \\ \Phi_u \end{bmatrix} - \begin{bmatrix} \mathbf{R} \\ \mathbf{M} \end{bmatrix} \left(\begin{bmatrix} zI - A & -B \end{bmatrix} \begin{bmatrix} \mathbf{R} \\ \mathbf{M} \end{bmatrix} \right)^{-1} \right\| \tag{3.16}$$

Due to the dependence of $\mathbf{\Delta}_c$ on $\mathbf{R}$ and $\mathbf{M}$, this expression is not convex with respect to the implementation matrices. Instead of directly including this expression as an optimization objective, we can penalize a corresponding heuristic

$$\left\| \begin{bmatrix} \Phi_x \\ \Phi_u \end{bmatrix} \begin{bmatrix} zI - A & -B \end{bmatrix} \begin{bmatrix} \mathbf{R} \\ \mathbf{M} \end{bmatrix} - \begin{bmatrix} \mathbf{R} \\ \mathbf{M} \end{bmatrix} \right\| \tag{3.17}$$

which is the equation error for (3.7) and is indeed convex with respect to the implementation matrices.

We are now ready to flesh out the second step of the two-step procedure. Given some desired closed-loop response (Φ_x, Φ_u), we solve the following optimization problem to obtain implementation matrices $(\mathbf{R}, \mathbf{M})$

$$\min_{\mathbf{R},\mathbf{M}} \left\| \begin{bmatrix} \Phi_x \\ \Phi_u \end{bmatrix} \begin{bmatrix} zI - A & -B \end{bmatrix} \begin{bmatrix} \mathbf{R} \\ \mathbf{M} \end{bmatrix} - \begin{bmatrix} \mathbf{R} \\ \mathbf{M} \end{bmatrix} \right\| + \lambda \left\| \begin{bmatrix} zI - A & -B \end{bmatrix} \begin{bmatrix} \mathbf{R} \\ \mathbf{M} \end{bmatrix} - I \right\| \tag{3.18a}$$

$$\text{s.t.} \quad \mathbf{R} \in \mathcal{P}_R, \mathbf{M} \in \mathcal{P}_M \tag{3.18b}$$

The first term in the objective is a heuristic for how close the implemented closed-loop responses are to the desired closed-loop responses. The second term in the objective is a stability-promoting objective, weighted by positive constant λ. For the control theorist who wishes to *enforce* stability rather than promote it, we may replace this stability-promoting objective with a constraint that the term is strictly less than 1 [4]. $\mathcal{P}_R$ and $\mathcal{P}_M$ represent constraint sets which may include sparsity and FIR constraints. Additionally, we could also include additional objectives such as $\mathcal{L}_1$ regularization to promote sparsity on the implementation matrices.

Optimization (3.18) can be decomposed into smaller subproblems to be solved in parallel, if norms are appropriately chosen (see [3] for examples). Thus, this optimization enjoys similar scalability benefits as those of standard SLS [3].

3.5 Closed-Loop vs. Controller Constraints

In this section, we discuss the physical interpretation of separately applying locality and delay constraints to the closed-loop and to the controller, and when such constraints are appropriate. This separation is not possible in standard SLS, since the closed-loop responses coincide with the implementation matrices for the controller.

First, we present a result on how applying controller constraints to closed-loop responses can be overly restrictive:

Proposition 3.3. Let $\mathbf{K}$ be the controller corresponding to the closed-loop responses $(\mathbf{\Phi}_x, \mathbf{\Phi}_u)$. Then, the operator $\mathbf{\Phi}_u$ lies in the range of the operator $\mathbf{K}$.

Proof. By Lemma 3.1, we have that $\mathbf{K}\mathbf{\Phi}_x = \mathbf{\Phi}_u$. $\qquad\qquad\square$

Proposition 3.3 shows that sparsity constraints (e.g. locality, delay) on $\mathbf{K}$ will always translate to sparsity constraints on $\mathbf{\Phi}_u$, but not $\mathbf{\Phi}_x$; directly applying these constraints on $\mathbf{\Phi}_x$ as well is overly restrictive.

Locality

Let $\mathcal{N}(i)$ denote some local neighborhood of subsystem i. Locality constraints restrict spectral components of $\mathbf{R}$ and $\mathbf{M}$ (or $\mathbf{\Phi}_x$ and $\mathbf{\Phi}_u$) to be zero at certain locations, i.e.

$$
\begin{aligned}
[R(k)]_{i,j} &= 0 \quad \forall j \notin \mathcal{N}(i) \\
[M(k)]_{i,j} &= 0 \quad \forall j \notin \mathcal{N}(i)
\end{aligned}
\tag{3.19}
$$

As an example, for a system with subsystems arranged in a chain configuration, with one state and input per subsystem and $\mathcal{N}(i)$ equal to the l closest neighbours of subsystem i, locality constraints result in banded diagonal $R(k)$ and $M(k)$ with a band width of $2l + 1 \ \forall k$.

When we apply locality constraints on implementation matrices as per (3.19), we enforce that subsystem i will only communicate with subsystems in $\mathcal{N}(i)$ for all time. When we apply locality constraints on the closed-loop responses (i.e. replace R and M in (3.19) with $\mathbf{\Phi}_x$ and $\mathbf{\Phi}_u$), we limit how far a disturbance at a subsystem spreads before it is contained. While both are useful, they are semantically distinct; there are situations when we want to enforce one but not the other, and the proposed two-step procedure provides a method to do so.

Delay

Let $d(i, j)$ denote the delay from subsystem j to subsystem i, which is generally proportional to the distance between subsystems i and j. Delay constraints are enforced as

$$[R(k)]_{i,j} = 0 \quad \forall k < d(i, j)$$
$$[M(k)]_{i,j} = 0 \quad \forall k < d(i, j) \tag{3.20}$$

As an example, for a system with subsystems arranged in a chain configuration, with one state and input per subsystem and $d(i, j)$ proportional to inter-subsystem distance, delay constraints result in banded diagonal $R(k)$ and $M(k)$, with wider bands for higher values of k.

When we apply delay constraints on the implementation matrices as per (3.20), we are ensuring that local controllers do not require information that cannot be communicated to them in time. For example, subsystem i cannot use any information about subsystem j that is more recent than $t - d(i, j)$. When we apply delay constraints on the closed-loop responses (i.e. replace R and M in (3.20) with Φ_x and Φ_u), we limit how fast a disturbance at subsystem j propagates to the state and input at subsystem i — this serves no clear purpose. Thus, delay constraints are only appropriate for implementation matrices, and should not be enforced on the closed-loop responses.

Delay and locality as optimization objectives

We can augment the objective in (3.18) with the following terms to encourage tolerance for communication delay

$$\sum_{k=1}^{T_c} \sum_i \sum_j e^{dist(i,j)-k} (\| [R(k)]_{i,j} \| + \| [M(k)]_{i,j} \|) \tag{3.21}$$

where $dist(i, j)$ is the distance between subsystems i and j in the network.

We can encourage tolerance for communication locality by using similar terms; note the removal of k from the exponential weight

$$\sum_{k=1}^{T_c} \sum_i \sum_j e^{dist(i,j)} (\| [R(k)]_{i,j} \| + \| [M(k)]_{i,j} \|) \tag{3.22}$$

Again taking the chain configuration with one state and one input per subsystem as an example, these terms encourage banded-diagonal $R(k)$ and $M(k)$ with higher penalties on elements farther away from the diagonal. Elements that survive despite heavy penalty represent edges in the network that require fast communication in order to best preserve the desired closed-loop response.

3.6 Simulations

Code required to replicate these simulations can be found in the SLS-MATLAB toolbox [6].

We use a system with 10 subsystems arranged in a chain configuration, with three actuators, which has the following A and B matrices

$$
A = \begin{bmatrix} 0.6 & 0.4 & 0 & \cdots & \\ 0.4 & 0.2 & \ddots & & \\ 0 & \ddots & \ddots & \ddots & \\ \vdots & & \ddots & 0.2 & 0.4 \\ & & & 0.4 & 0.6 \end{bmatrix}, \quad (B)_{i,j} = \begin{cases} 1 & \text{if } (i,j) \in ((3,1),(6,2),(10,3)) \\ 0 & \text{otherwise} \end{cases}
$$

$$(3.23)$$

The system is marginally stable, with a spectral radius of 1.

Localized LQR controller

We provide an example in which the proposed two-step procedure substantially outperforms both standard SLS and virtually local SLS [4].

The goal is to synthesize a controller with some standard LQR cost and communication constraints — each sub-controller is only allowed to use information from its two neighbouring sub-controllers, and communication speed between sub-controllers is restricted to be the same speed as the propagation of the discrete-time dynamics. We use standard SLS, virtually local SLS, and two-step SLS (the technique proposed in this chapter) with horizon 20 to solve this problem, and the results are shown in Table 3.1. For each controller, we include the LQR cost normalized by the optimal, non-communication-constrained LQR controller, as well as the spectral radius of the internal dynamics.

The application of the desired constraints renders the standard SLS problem infeasible, as the desired constraints overly restrict the closed-loop responses. Using virtually local SLS, we are able to obtain a controller that meets the requirements. However, this controller performs nearly 30% worse than the optimal unconstrained controller. The controller is internally stable, with an internal spectral radius of 0.847.

Table 3.1: Comparison of LQR costs

Controller	LQR cost	Internal spectral radius
Standard SLS, $T = 20$		*Infeasible*
Virtually local SLS, $T = 20$	1.294	0.847
Two-step SLS, $T_c = 20$	1.033	0.876
Two-step SLS, $T_c = 2$	1.034	0.851

To apply the two-step procedure, we first synthesize a centralized FIR controller with a horizon of 20 timesteps using SLS. Then, we find appropriate implementation matrices using (3.18). We first search over implementation matrices with controller order $T_c = 20$. This yields a controller that performs only 3% worse than the optimal unconstrained controller, and is internally stable, with an internal spectral radius that is slightly higher than that of virtually local SLS. We also attempt to synthesize a controller with order $T_c = 2$ using the two-step procedure. Interesting, this lower-order controller performs almost as well as the full-order controller, with only 0.1% performance degradation. This suggests that in this case, highly delayed information (which correspond to higher order terms of the implementation matrices) are not very useful to the controller. Additionally, the lower-order controller has similar internal stability as the higher-order controller. Overall, in this example, both two-step controllers are vastly preferable compared to existing SLS techniques.

3.7 Conclusions and Future Work

By separating controller synthesis from closed-loop synthesis, we are able to apply constraints to the controller without unnecessarily limiting the closed-loop response. Our proposed two-step procedure offers an alternative approach for scenarios in which standard SLS is infeasible.

The theory in this chapter can be extended to further clarify the relationship between closed-loop responses and their corresponding implementation matrices. For a given network, techniques from this chapter can be used to explore which communication links in a system are crucial to optimal control performance, and which are not. Additionally, the example provided shows that in some cases, extremely low-order controllers perform nearly as well as high-order controllers — more investigation is required to determine the scenarios in which this order reduction is available. Furthermore, we are interested in comparing this technique with other distributed control techniques, particularly spatial truncations of dense LQR controllers.

Chapter 4

DISTRIBUTED STRUCTURED ROBUST CONTROL

[1]   J. S. Li and J. C. Doyle, "Distributed Robust Control for Systems with Structured Uncertainties," in *IEEE Conference on Decision and Control*, 2022, pp. 1702–1707. DOI: 10.1109/CDC51059.2022.9992622. [Online]. Available: http://arxiv.org/abs/2204.02493,

Overview: We present D-Φ iteration: an algorithm for distributed, localized, and scalable synthesis of robust controllers for systems with structured uncertainties. This algorithm combines the SLS parametrization for distributed control with stability criteria from $\mathcal{L}_1$, $\mathcal{L}_\infty$, and ν robust control. We show in simulation that the controller generated by this algorithm achieves good nominal performance while greatly increasing the robust stability margin compared to the LQR controller. To the best of our knowledge, this is the first distributed and localized algorithm for structured robust controller synthesis; furthermore, algorithm complexity depends only on the size of local neighborhoods and is independent of global system size. We additionally characterize the suitability of different robustness criteria for distributed and localized computation.

4.1   Introduction

Robust control theory [7]–[9] provides stability and performance guarantees in the face of model uncertainty, which arises from imprecise models or unexpected operating conditions. Robust control theory plays a crucial role in mitigating the undesirable effects of model uncertainty on various engineering systems.

For large-scale systems, distributed $\mathcal{H}_\infty$ robust control has been studied for linear spatially invariant systems [10] and linear symmetric systems [11], among others; more recently, distributed $\mathcal{L}_1$ robust controller synthesis for arbitrary linear systems is proposed in [12]. However, to the best of our knowledge, no distributed methods address *structured* uncertainty, which often arises in large systems; for instance, in a power grid, we may have parametric uncertainty on electrical properties of transmission lines between buses, but no uncertainty between unconnected buses — this imposes structure on system uncertainty. Incorporating knowledge of this structure into the synthesis procedure allows control engineers to provide robust stability margins with minimal conservatism, which has been historically important in the design of aerospace systems.

In this chapter, we leverage the SLS parametrization to provide distributed synthesis methods for structured robust control. We consider diagonal time-varying uncertainty and use $\mathcal{L}_1$, $\mathcal{L}_\infty$, and ν [13] robustness criteria to formulate synthesis problems. We also present D-Φ iteration, an algorithm for distributed synthesis of robust controllers. To the best of our knowledge, this is the first distributed and localized algorithm for structured robust control.

We first describe the SLS formulation as a generic optimization problem, and how it can be used to describe state feedback, full control, and output feedback problems. Then, we describe the concept *separability*, extending concepts from system level synthesis [3], and discuss and how separability relates to scalability and distributed computation. Then, we describe how $\mathcal{L}_1$, $\mathcal{L}_\infty$, and ν robustness criteria translate into separable objectives and constraints on the control problem, and present two versions of the scalable D-Φ iteration algorithm for robust control. The efficacy of the algorithms is demonstrated via simulations; D-Φ iteration provides good robust stability margins while mostly preserving nominal performance.

4.2 System Level Synthesis for Linear Control Problems

Using SLS, we can cast any linear control problem as an optimization of the form

$$\min_{\Phi} \quad f(\Phi) \quad \text{s.t.} \quad \Phi \in \mathcal{S}_a \cap \mathcal{P} \tag{4.1}$$

where f is some convex functional, and $\mathcal{S}_a$ and $\mathcal{P}$ are convex sets. $\mathcal{S}_a$ represents the *achievability constraint*, which is enforced in all SLS problems; it ensures that we only search over closed-loop responses Φ which can be achieved using a causal, internally stable controller. This is made mathematically explicit in Theorem 2.1. $\mathcal{P}$ represents additional, optional constraints; we typically use $\mathcal{P}$ to enforce sparsity on Φ, corresponding to local communication, delayed communication, or local disturbance rejection. In Chapter II, we described how state feedback control can be cast in the form of (4.1) — in this case, the decision variable is $\Phi = \begin{bmatrix} \Phi_x \\ \Phi_u \end{bmatrix}$, where Φ_x and Φ_u map from exogenous disturbance to state and input, respectively. The *state feedback achievability constraint* is given by (2.4). We now provide definitions of Φ and $\mathcal{S}_a$ for the full control and output feedback cases.

The full control problem can be formulated in the frequency domain as

$$z\mathbf{x} = A\mathbf{x} + \mathbf{u} + \mathbf{w} \tag{4.2a}$$

$$\mathbf{y} = C\mathbf{x} + \mathbf{v} \tag{4.2b}$$

where $\mathbf{v}$ represents measurement noise. Apply linear causal controller $\mathbf{u} = \mathbf{L}\mathbf{y}$ for some transfer matrix $\mathbf{L}$. We remark that full control (i.e. full actuation, sparse sensing) is the dual of state feedback (i.e. full sensing, sparse actuation). Define $\mathbf{\Phi}_w$ and $\mathbf{\Phi}_v$, closed-loop responses from disturbance and measurement noise to state

$$\mathbf{x} = (zI - A - LC)^{-1}\mathbf{w} + (zI - A - LC)^{-1}L\mathbf{v}$$
$$= \begin{bmatrix} \mathbf{\Phi}_w & \mathbf{\Phi}_v \end{bmatrix} \begin{bmatrix} \mathbf{w} \\ \mathbf{v} \end{bmatrix} \tag{4.3}$$

Results from state feedback apply by duality. The *full control achievability constraint* is

$$\begin{bmatrix} \mathbf{\Phi}_w & \mathbf{\Phi}_v \end{bmatrix} \begin{bmatrix} zI - A \\ -C \end{bmatrix} = I, \quad \mathbf{\Phi}_w, \mathbf{\Phi}_v \in \frac{1}{z}\mathcal{RH}_\infty \tag{4.4}$$

Any full control problem can be cast as (4.1), where $\mathbf{\Phi} = \begin{bmatrix} \mathbf{\Phi}_w & \mathbf{\Phi}_v \end{bmatrix}$, and $\mathbf{\Phi} \in \mathcal{S}_a$ is equivalent to $\mathbf{\Phi}$ satisfying (4.4).

For the output feedback problem defined by (2.2) and (4.2b), i.e.

$$z\mathbf{x} = A\mathbf{x} + B\mathbf{u} + \mathbf{w}$$
$$\mathbf{y} = C\mathbf{x} + \mathbf{v} \tag{4.5}$$

with linear causal controller $\mathbf{K}$ and observer $\mathbf{L}$, we define four closed-loop responses mapping disturbance and measurement noise to state and control: $\mathbf{\Phi}_{xw}$, $\mathbf{\Phi}_{xv}$, $\mathbf{\Phi}_{uw}$, $\mathbf{\Phi}_{uv}$, such that

$$\begin{bmatrix} \mathbf{x} \\ \mathbf{u} \end{bmatrix} = \begin{bmatrix} \mathbf{\Phi}_{xw} & \mathbf{\Phi}_{xv} \\ \mathbf{\Phi}_{uw} & \mathbf{\Phi}_{uv} \end{bmatrix} \begin{bmatrix} \mathbf{w} \\ \mathbf{v} \end{bmatrix} \tag{4.6}$$

The *output feedback achievability constraint* resembles a combination of the state feedback and full control achievability constraints (2.4) and (4.4)

$$\begin{bmatrix} zI - A & -B \end{bmatrix} \begin{bmatrix} \mathbf{\Phi}_{xw} & \mathbf{\Phi}_{xv} \\ \mathbf{\Phi}_{uw} & \mathbf{\Phi}_{uv} \end{bmatrix} = \begin{bmatrix} I & 0 \end{bmatrix} \tag{4.7a}$$

$$\begin{bmatrix} \mathbf{\Phi}_{xw} & \mathbf{\Phi}_{xv} \\ \mathbf{\Phi}_{uw} & \mathbf{\Phi}_{uv} \end{bmatrix} \begin{bmatrix} zI - A \\ -C \end{bmatrix} = \begin{bmatrix} I \\ 0 \end{bmatrix} \tag{4.7b}$$

$$\mathbf{\Phi}_{xw}, \mathbf{\Phi}_{uw}, \mathbf{\Phi}_{xv} \in \frac{1}{z}\mathcal{RH}_\infty, \mathbf{\Phi}_{uv} \in \mathcal{RH}_\infty \tag{4.7c}$$

Detailed derivations can be found in Section 5 of [3]. Any output feedback problem can be cast as (4.1), where $\mathbf{\Phi} = \begin{bmatrix} \mathbf{\Phi}_{xw} & \mathbf{\Phi}_{xv} \\ \mathbf{\Phi}_{uw} & \mathbf{\Phi}_{uv} \end{bmatrix}$, and $\mathbf{\Phi} \in \mathcal{S}_a$ is equivalent to $\mathbf{\Phi}$ satisfying (4.7).

4.3 Separability and Computation

We present definitions of separability and compatibility, extending concepts from [3]. We describe how separable objectives and constraints in (4.1) translate to distributed computation for both optimal and robust control.

Separability

Definition 4.1. Define transfer matrices $\Phi = \arg\min f(\Phi)$ and Ψ, where $(\Psi)_{i,j} = \arg\min f_{ij}((\Psi)_{i,j})$ and f, f_{ij} are some functionals. f is an *element-wise separable functional* if there exist sub-functionals f_{ij} such that $f(\Phi) = f(\Psi)$.

Relevant element-wise separable functionals include $\|\cdot\|_{\mathcal{H}_2}$ and $\|\cdot\|_{1 \to \infty}$ (i.e. maximum absolute element). The latter is the norm associated with ν robustness [13].

Definition 4.2. Let $\mathcal{P}$ represent some constraint set. $\mathcal{P}$ is an *element-wise separable constraint* if there exist sub-constraint sets $\mathcal{P}_{ij}$ such that $\Phi \in \mathcal{P} \Leftrightarrow (\Phi)_{i,j} \in \mathcal{P}_{ij}$.

Sparsity constraints on Φ are element-wise separable.

Definition 4.3. Define transfer matrices $\Phi = \arg\min f(\Phi)$ and Ψ, where $(\Psi)_{i,:} = \arg\min f_i((\Psi)_{i,:})$ and f, f_i are some functionals. f is a *row-separable functional* if there exist sub-functionals f_i such that $f(\Phi) = f(\Psi)$.

Definition 4.4. Let $\mathcal{P}$ represent some constraint set. $\mathcal{P}$ is a *row separable constraint* if there exist sub-constraint sets $\mathcal{P}_i$ such that $\Phi \in \mathcal{P} \Leftrightarrow (\Phi)_{i,:} \in \mathcal{P}_i$.

Column separable functionals and constraints are defined analogously. Clearly, any element-wise separable functional or constraint is also both row and column separable. A relevant row separable functional is $\|\cdot\|_{\infty \to \infty}$ (i.e. maximum absolute row sum), which is the norm associated with $\mathcal{L}_1$ robustness. A relevant column separable functional is $\|\cdot\|_{1 \to 1}$ (i.e. maximum absolute column sum), which is the norm associated with $\mathcal{L}_\infty$ robustness. We also see that constraints of the form $G\Phi = H$ are column separable, while constraints of the form $\Phi G = H$ are row separable. Thus, the state feedback achievability constraint is column separable, and the full control achievability constraint is row separable. For the output feedback achievability constraints, (4.7a) is column separable, (4.7b) is row separable, and (4.7c) is element-wise separable.

Definition 4.5. Optimization problem (4.1) is a *fully row separable optimization* when all objectives and constraints are row separable. It is a *fully column separable optimization* when all objectives and constraints are column separable.

Definition 4.6. Optimization problem (4.1) is a *partially separable optimization* when all objectives and constraints are either row or column separable, but the overall problem is not fully separable.

The separability of the SLS state feedback, full control, and output feedback problems are inherently limited by their achievability constraints, since these must always be enforced. The state feedback problem can be fully column separable, but not fully row separable; the opposite is true for the full control problem[1]. The output feedback problem cannot be fully separable, since it includes a mix of column and row separable achievability constraints. We now describe the computational implications of separability.

Scalability and computation

A fully separable optimization is easily solved via distributed computation. For instance, if problem (4.1) is fully row separable, we can solve the subproblem

$$\min_{(\mathbf{\Phi})_{i,:}} \quad f_i((\mathbf{\Phi})_{i,:}) \quad \text{s.t.} \quad (\mathbf{\Phi})_{i,:} \in \mathcal{S}_{a_i} \cap \mathcal{P}_i \tag{4.8}$$

at each subsystem i in the system, in parallel.

We often enforce sparsity on $\mathbf{\Phi}$ to constrain inter-subsystem communication to local neighborhoods of size d. This translates to $(\mathbf{\Phi})_{i,j} = 0 \quad \forall j \notin \mathcal{N}_d(i)$, where $\mathcal{N}_d(i)$ is the set of subsystems that are in subsystem i's local neighborhood, and $|\mathcal{N}_d(i)|$ depends on d. Then, the size of the decision variable in subproblem (4.8) scales with d. Each subsystem solves a subproblem in parallel; overall, computational complexity scales with d instead of system size N. This is highly beneficial for large systems, where we can choose d much smaller than N.

For partially separable problems, we apply the alternating direction method of multipliers (ADMM) [14]. We first rewrite (4.1) in terms of row separable and column separable objectives and constraints

$$\min_{\mathbf{\Phi}} \quad f^{(\text{row})}(\mathbf{\Phi}) + f^{(\text{col})}(\mathbf{\Phi})$$
$$\text{s.t.} \quad \mathbf{\Phi} \in \mathcal{S}_a^{(\text{row})} \cap \mathcal{S}_a^{(\text{col})} \cap \mathcal{P}^{(\text{row})} \cap \mathcal{P}^{(\text{col})} \tag{4.9}$$

---

[1]The only exception is if all system matrices are diagonal; in this case, all achievability constraints are element-wise separable.

Note that an element-wise separable objective can appear in both $f^{\text{(row)}}$ and $f^{\text{(col)}}$; similarly, an element-wise separable constraint $\mathcal{P}$ can appear in both $\mathcal{P}^{\text{(row)}}$ and $\mathcal{P}^{\text{(col)}}$. We now introduce duplicate variable Ψ and dual variable Λ. The ADMM algorithm is iterative; for each iteration k, we perform the following computations

$$\Phi^{k+1} = \underset{\Phi}{\text{argmin}} \quad f^{\text{(row)}}(\Phi) + \frac{\gamma}{2}\|\Phi - \Psi^k + \Lambda^k\|_F \tag{4.10a}$$
$$\text{s.t.} \quad \Phi \in \mathcal{S}_a^{\text{(row)}} \cap \mathcal{P}^{\text{(row)}}$$

$$\Psi^{k+1} = \underset{\Psi}{\text{argmin}} \quad f^{\text{(col)}}(\Psi) + \frac{\gamma}{2}\|\Phi^{k+1} - \Psi + \Lambda^k\|_F \tag{4.10b}$$
$$\text{s.t.} \quad \Psi \in \mathcal{S}_a^{\text{(col)}} \cap \mathcal{P}^{\text{(col)}}$$

$$\Lambda^{k+1} = \Lambda^k + \Phi^{k+1} - \Psi^{k+1} \tag{4.10c}$$

ADMM separates (4.1) into a row separable problem (4.10a) and a column separable problem (4.10b), and encourages consensus between Φ and Ψ (i.e. $\Phi = \Psi$) via the γ-weighted objective and (4.10c). When both $\|\Phi^{k+1} - \Psi^{k+1}\|_F$ and $\|\Phi^{k+1} - \Phi^k\|_F$ are sufficiently small, the algorithm converges; Φ^k is the optimal solution to (4.1).

Optimizations (4.10a) and (4.10b) are fully separable; as described above, they enjoy complexity that scales with local neighborhood size instead of global system size. Additionally, due to sparsity constraints on Φ and Ψ, only local communication is required between successive iterations. Thus, partially separable problems also enjoy computational complexity that scale with local neighborhood size instead of global system size [3], [15]. However, partially separable problems require iterations, making them more computationally complex than fully separable problems. The separability of objectives and constraints in (4.1) can also affect convergence rate, as we will describe next.

Definition 4.7. Let $\mathfrak{R}, \mathfrak{C} \in \mathbb{Z}^+$ indicate some sets of indices. Define f_{sub}, a functional of $(\Phi)_{\mathfrak{R},\mathfrak{C}}$, and f, a functional of Φ. Define transfer matrices Φ and Ψ, and let $(\Phi)_{i,j} = (\Psi)_{i,j} \ \ \forall (i,j) \notin (\mathfrak{R}, \mathfrak{C})$. f_{sub} is a *compatible sub-functional* of functional f if minimizing f_{sub} is compatible with minimizing f, i.e. $f_{\text{sub}}((\Phi)_{\mathfrak{R},\mathfrak{C}}) \leq f_{\text{sub}}((\Psi)_{\mathfrak{R},\mathfrak{C}}) \Rightarrow f(\Phi) \leq f(\Psi)$.

Definition 4.8. Let $\mathfrak{R}, \mathfrak{C} \in \mathbb{Z}^+$ indicate some sets of indices. Constraint set $\mathcal{P}_{\text{sub}}$ is a *compatible sub-constraint* of constraint set $\mathcal{P}$ if $\Phi \in \mathcal{P} \Rightarrow (\Phi)_{\mathfrak{R},\mathfrak{C}} \in \mathcal{P}_{\text{sub}}$ for some set of elements $(\Phi)_{\mathfrak{R},\mathfrak{C}}$.

Compatibility complements separability. If functional f is separable into sub-functionals f_i, then each f_i is a compatible sub-functional of f; similar arguments apply for constraints.

Definition 4.9. Let $\mathfrak{R}$, $\mathfrak{C} \in \mathbb{Z}^+$ indicate some sets of indices. The partially separable problem (4.9) can be decomposed into row and column sub-functionals and sub-constraints. We say (4.9) is a *balanced ADMM* problem if, for any set of elements $(\Phi)_{\mathfrak{R},\mathfrak{C}}$ which appear together in a sub-constraint, there exists a matching sub-functional f_{sub} that is compatible with f, and depends only on $(\Phi)_{\mathfrak{R},\mathfrak{C}}$.

Intuitively, a balanced partially separable problem converges faster than an unbalanced one. For example, consider an output feedback problem with a row separable objective. This is an unbalanced partially separable problem; though all row sub-constraints have a matching sub-objective, none of the column sub-constraints have matching sub-objectives. Thus, we are only able to minimize the objective in the row computation (4.10a); this results in slow convergence, and places more burden on consensus between Φ and Ψ than a balanced problem would. More unbalanced examples include a state feedback problem with a row separable objective, or a full control problem with a column separable objective. To balance an output feedback problem, we require an element-wise separable objective f.

To summarize: for both fully separable and partially separable problems, computational complexity scales independently of system size. However, partially separable problems require iteration, while fully separable problems do not. For partially separable problems, we prefer a balanced problem to an unbalanced problem due to faster convergence. Thus, element-wise separability (e.g. $\mathcal{H}_2$ optimal control, ν robustness) is desirable for two reasons: firstly, for state feedback and full control, element-wise separable objectives give rise to fully separable problems. Secondly, for output feedback, where ADMM iterations are unavoidable, element-wise separable objectives give rise to balanced problems. Finally, we remark that $\mathcal{H}_\infty$ robust control problems are not at all separable, and make for highly un-scalable computations; this motivates our use of $\mathcal{L}_1$, $\mathcal{L}_\infty$, and ν robustness.

4.4 Robust Stability

We present distributed algorithms for structured robust control, using criterion from $\mathcal{L}_1$, $\mathcal{L}_\infty$, and ν robustness. We consider diagonal nonlinear time-varying (DNLTV) uncertainties Δ and strictly causal LTI closed-loops Φ. We leverage the

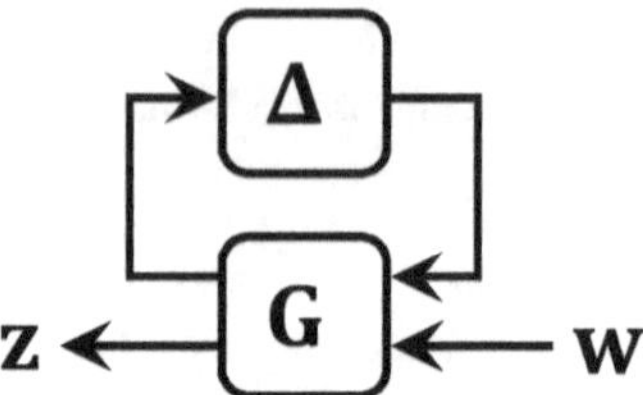

Figure 4.1: Feedback interconnection of transfer matrix **G** and uncertainty **Δ**. **G** is the nominal closed-loop response from disturbance **w** to regulated output **z**.

SLS formulation to formulate robust control problems as distributed optimization problems.

Robust stability conditions

Let transfer matrix **G** map disturbance **w** to regulated output **z**. Generally, **z** is a linear function of state **x** and input **u**, i.e. $\mathbf{G} = H\mathbf{\Phi}$ for some constant matrix H. Thus, **G** is strictly casual and LTI. Assume that we have some fixed closed-loop map **Φ**, and therefore fixed **G**. We construct positive constant magnitude matrix $M = \sum_{p=1}^{\infty} |\mathbf{G}(p)|$, where $\mathbf{G}(p)$ are spectral elements of **G**, and $|\cdot|$ denotes the element-wise absolute value. Let $\mathcal{D}$ be the set of positive diagonal matrices

$$\mathcal{D} = \{D \in \mathbb{R}^{n \times n} : (D)_{i,j} = 0 \quad \forall i \neq j, (D)_{i,i} > 0 \quad \forall i\} \tag{4.11}$$

Lemma 4.1. The interconnection in Figure 4.1 is robustly stable in the $\mathcal{L}_1$ sense for all DNLTV **Δ** such that $\|\mathbf{\Delta}\|_{\infty \to \infty} < \frac{1}{\beta}$ if and only if $\inf_{D \in \mathcal{D}} \|DMD^{-1}\|_{\infty \to \infty} \leq \beta$. Proof in [7].

Lemma 4.2. The interconnection in Figure 4.1 is robustly stable in the $\mathcal{L}_\infty$ sense for all DNLTV **Δ** such that $\|\mathbf{\Delta}\|_{1 \to 1} < \frac{1}{\beta}$ if and only if $\inf_{D \in \mathcal{D}} \|DMD^{-1}\|_{1 \to 1} \leq \beta$. Proof: equivalent to applying Lemma 4.1 to $M^\top$.

Lemma 4.3. The interconnection in Figure 4.1 is robustly stable in the ν sense for all DNLTV **Δ** such that $\|\mathbf{\Delta}\|_{\infty \to 1} < \frac{1}{\beta}$ if $\inf_{D \in \mathcal{D}} \|DMD^{-1}\|_{1 \to \infty} \leq \beta$. Additionally, if $\exists D \in \mathcal{D}$ s.t. DMD^{-1} is diagonally maximal, then this condition is both sufficient *and* necessary. Proof: Theorem 4 in [13].

Definition 4.10. Matrix $A \in \mathbb{R}^{n \times n}$ is *diagonally maximal* if $\exists k$ s.t. $|(A)_{k,k}| = \max_{i,j} |(A)_{i,j}|$, i.e. the maximum element in A lies on its diagonal.

In general, computing the $\|\cdot\|_{\infty\to1}$ norm is NP-hard; for diagonal Δ, $\|\Delta\|_{\infty\to1} = \sum_k |(\Delta)_{k,k}|$ [13].

Let the nominal performance be $\|Q\Phi\|_{\text{perf}}$ for some norm $\|\cdot\|_{\text{perf}}$ and some constant matrix Q. Then, leveraging the robust analysis results from Lemmas 4.1-4.3, we can pose the *synthesis* problem for nominal performance and robust stability as

$$\min_{\Phi,M,D} \quad \|Q\Phi\|_{\text{perf}} + \|DMD^{-1}\|_{\text{stab}}$$

$$\text{s.t.} \quad M = \sum_{p=1}^{T} |H\Phi(p)|, \quad \Phi \in \mathcal{S}_a \cap \mathcal{P}, \quad D \in \mathcal{D} \tag{4.12}$$

where for ease of computation we assume that Φ is finite impulse response (FIR) with horizon T. We intentionally leave norms ambiguous; $\|\cdot\|_{\text{stab}}$ can be $\|\cdot\|_{\infty\to\infty}$, $\|\cdot\|_{1\to1}$, or $\|\cdot\|_{1\to\infty}$ for $\mathcal{L}_1$, $\mathcal{L}_\infty$, and ν robust stability, respectively. $\|DMD^{-1}\|$ corresponds to the robust stability margin $\frac{1}{\beta}$; robust stability is guaranteed for all Δ such that $\|\Delta\| < \frac{1}{\beta}$, for the appropriate norm on Δ. Smaller β corresponds to stability guarantees for a larger set of Δ. The nominal performance norm $\|\cdot\|_{\text{perf}}$ may be different from $\|\cdot\|_{\text{stab}}$.

D-Φ iteration

Problem (4.12) is nonconvex, and does not admit a convex reformulation. Inspired by the D-K iteration method from $\mathcal{H}_\infty$ robust control [8], we adopt an iterative approach. We heuristically minimize (4.12) by iteratively fixing D and optimizing over Φ in the "Φ step", then fixing Φ and optimizing (or randomizing) over D in the "D step".

We remark that problem (4.12) poses both nominal performance and robust stability as objectives. If we already know the desired robust stability margin $\beta_{\max}^{-1}$, we can omit the stability objective and instead enforce a constraint $\|DMD^{-1}\|_{\text{stab}} \leq \beta_{\max}$, as is done in D-K iteration; similarly, if we already know the desired nominal performance α, we can omit the performance objective and enforce $\|Q\Phi\|_{\text{perf}} \leq \alpha$.

The Φ step solves the following problem

$$\Phi, M = \operatorname*{argmin}_{\Phi,M} \quad \|Q\Phi\|_{\text{perf}}$$

$$\text{s.t.} \quad \|DMD^{-1}\|_{\text{stab}} \leq \beta$$

$$M = \sum_{p=1}^{T} |H\Phi(p)|, \quad \Phi \in \mathcal{S}_a \cap \mathcal{P} \tag{4.13}$$

for some fixed value β and scaling matrix $D \in \mathcal{D}$. The separability of problem (4.13) is characterized by the separability of its objective and constraints. As previously mentioned, $\mathcal{P}$ typically consists of sparsity constraints on Φ; this constraint is element-wise separable. The separability of other constraints and objectives in (4.13) are as follows:

- If Q is separably diagonal, $\|Q\Phi\|_{\mathrm{perf}}$ has the same separability as $\|\cdot\|_{\mathrm{perf}}$; If not, it is column separable if and only if $\|\cdot\|_{\mathrm{perf}}$ is column separable.

- $\|DMD^{-1}\|_{\mathrm{stab}} < \beta$ has the same separability as $\|\cdot\|_{\mathrm{stab}}$.

- If H is separably diagonal, $M = \sum_{p=1}^{T} |H\Phi(p)|$ is element-wise separable; if not, it is column separable.

- $\Phi \in \mathcal{S}_a$ is column separable for state feedback, row separable for full control, and partially separable for output feedback.

Definition 4.11. For state feedback, the product $Q\Phi$ may be written as $Q_x\Phi_x + Q_u\Phi_u$ for some matrices Q_x and Q_u. Q is *separably diagonal* if both Q_x and Q_u are diagonal matrices. Analogous definitions apply to the full control and output feedback case.

Table 4.1 summarizes the separability of (4.13) for state feedback, full control, and output feedback problems with $\mathcal{H}_\infty$, $\mathcal{L}_1$, $\mathcal{L}_\infty$, and ν robustness, where we assume that Q and H are separably diagonal and $\|\cdot\|_{\mathrm{perf}}$ has the same type of separability as $\|\cdot\|_{\mathrm{stab}}$. Note that $\mathcal{H}_\infty$ is not separable for any problem. For state feedback, $\mathcal{L}_\infty$ and ν are the preferred stability criteria; for full control, $\mathcal{L}_1$ and ν are the preferred stability criteria. For output feedback, ν is the only criterion that produces a balanced partially separable problem. Overall, ν robustness is preferable in all three cases, resulting in either fully separable formulations that require no iterations, or balanced partially separable formulations, which have preferable convergence properties. Though convergence properties vary, the Φ step (4.13) can be computed with complexity that scales with local neighborhood size d instead of global system size for all non-$\mathcal{H}_\infty$ cases.

The D step solves the following problem

$$D = \underset{D}{\mathrm{argmin}} \quad \|DMD^{-1}\|_{\mathrm{stab}} \quad \text{s.t.} \quad D \in \mathcal{D} \qquad (4.14)$$

Table 4.1: Separability of Φ step of D-Φ iteration

	State Feedback	Full Control	Output Feedback
$\mathcal{H}_\infty$	No	No	No
$\mathcal{L}_1$	Partial, Unbalanced	Full	Partial, Unbalanced
$\mathcal{L}_\infty$	Full	Partial, Unbalanced	Partial, Unbalanced
ν	Full	Full	Partial, Balanced

for some fixed magnitude matrix M. For ν robustness, the minimizing D step (4.14) can be recast as a linear program

$$l_i, \eta = \operatorname*{argmin}_{l_i, \eta} \quad \eta \quad \text{s.t.} \quad \log(M)_{i,j} + l_i - l_j \leq \eta$$
$$\forall (M)_{i,j} \neq 0, \quad 1 \leq i, j \leq n \tag{4.15}$$

The optimal solution D can be recovered as $D = \operatorname{diag}(\exp(l_1), \exp(l_2), \ldots \exp(l_n))$. Problem (4.15) can be distributedly computed using ADMM consensus [14]. Let $x_i = \begin{bmatrix} \eta_i \\ L_{j@i} \end{bmatrix}$ be the variable at subsystem i, where η_i is subsystem i's value of η, and $L_{j@i}$ is a vector containing $l_{j@i}$: subsystem i's values of l_j for all $j \in \mathcal{N}_d(i)$. The goal is for all subsystems to reach consensus on η and $l_i, 1 \leq i \leq N$. We introduce dual variable y_i and averaging variable $\bar{x}_i = \begin{bmatrix} \bar{\eta}_i \\ \bar{L}_{j@i} \end{bmatrix}$. For each iteration k, subsystem i performs the following computations

$$x_i^{k+1} = \operatorname*{argmin}_{x_i} \quad \eta_i + (y_i^k)^\top (x_i - \bar{x}_i^k) + \gamma \|x_i - \bar{x}_i^k\|_2^2$$
$$\text{s.t.} \quad \forall j \in \mathcal{N}_d(i), (M)_{i,j} + l_{i@i} - l_{j@i} \leq \eta_i \tag{4.16a}$$

$$\bar{\eta}_i^{k+1} = \frac{1}{|\mathcal{N}_d(i)|} \sum_{j \in \mathcal{N}_d(i)} \eta_j$$
$$\bar{l}_i = \frac{1}{|\mathcal{N}_d(i)|} \sum_{j \in \mathcal{N}_d(i)} l_{i@j}, \quad \bar{l}_{j@i}^{k+1} = \bar{l}_j, \forall j \in \mathcal{N}_d(i) \tag{4.16b}$$

$$y_i^{k+1} = y_i^k + \frac{\gamma}{2}(x_i^{k+1} - \bar{x}_i^{k+1}) \tag{4.16c}$$

where γ is a user-determined parameter, and iterations stop when consensus is reached, i.e. differences between relevant variables are sufficiently small. The size of optimization variable x_i depends only on local neighborhood size d; thus, the complexity of (4.16a) scales independently of global system size. Computation (4.16b) requires communication, but only within the local neighborhood. Also, by the definition of Φ and the construction of M, $(M)_{i,j} = 0 \quad \forall j \notin \mathcal{N}_d(i)$. Thus, for a fully connected system, solving (4.16) is equivalent to solving (4.15). Additionally, consensus problems are balanced as per Definition 4.9, so (4.16) converges relatively quickly.

For $\mathcal{L}_1$ robustness, (4.14) is solved by setting $D = \text{diag}(v_1, v_2, \ldots v_n)^{-1}$, where v is the eigenvector corresponding to the largest-magnitude eigenvalue of M [7]. This computation does not lend itself to scalable distributed computation. To ameliorate this, we propose an alternative formulation that randomizes instead of minimizing over D. This can be written in the form of an optimization problem as

$$D = \underset{D}{\text{argmin}} \quad 0 \quad \text{s.t.} \quad \|DMD^{-1}\|_{\text{stab}} \le \beta \tag{4.17}$$

The randomizing formulation lends itself to distributed computation. Also, we remark that (4.14) can be solved by iteratively solving (4.17) to search for the lowest feasible value of β.

Define vectors

$$v = \begin{bmatrix} (D)_{1,1} \\ (D)_{2,2} \\ \vdots \\ (D)_{n,n} \end{bmatrix}, v^{-1} = \begin{bmatrix} (D)_{1,1}^{-1} \\ (D)_{2,2}^{-1} \\ \vdots \\ (D)_{n,n}^{-1} \end{bmatrix} \tag{4.18}$$

We can rewrite constraint $\|DMD^{-1}\|_{\text{stab}} \le \beta$ as $Mv^{-1} \le \beta v^{-1}$ for $\mathcal{L}_1$ stability, and $M^\top v \le \beta v$ for $\mathcal{L}_\infty$ stability. Then, problem (4.17) can be formulated as a scalable distributed ADMM consensus problem using similar techniques as (4.15).

Both versions of the D step ((4.14) and (4.17)) for D-Φ iteration are simpler than the D step in D-K iteration [8], which requires a somewhat involved frequency fitting process. Also, all separable versions of the proposed D step are less computationally intensive than the Φ step (4.13), since the decision variable in the D step is much smaller. Table 4.2 summarizes the scalability of different versions of the D step for different robustness criteria. "Minimize" refers to solving (4.14) directly; "Iteratively Minimize" refers to solving (4.14) by iteratively solving (4.17) to search for the lowest feasible β; "Randomize" refers to solving (4.14). $\checkmark$ indicates that we can use scalable distributed computation, and $\boldsymbol{X}$ indicates that no scalable distributed formulation is available; by scalable, we mean complexity that scales independently of global system size. For iterative minimization, there is the obvious caveat of iterations incurring additional computational time; however, for $\mathcal{L}_1$ and $\mathcal{L}_\infty$, iterative minimization is more scalable than direct minimization. Additionally, we show in

Table 4.2: Scalability of D step of D-Φ iteration

	Minimize	Iteratively Minimize	Randomize
$\mathcal{H}_\infty$	✗	✗	✗
$\mathcal{L}_1$	✗	✓	✓
$\mathcal{L}_\infty$	✗	✓	✓
ν	✓	✓	✓

the next section that algorithms using the randomizing D step perform similarly as algorithms using the minimizing D step; thus, iterative minimization may be unnecessary. Overall, ν robustness appears to be preferable for scalability purposes for both the Φ step and D step.

We now present two algorithms for D-Φ iteration. Algorithm 4.1 is based on minimizing over D (4.14), while Algorithm 4.2 is based on randomizing over D (4.17). Both algorithms compute the controller Φ which achieves optimal nominal performance for some desired robust stability margin $\beta_{\max}^{-1}$.

Algorithm 4.1 D-Φ iteration with minimizing D step

---

    input : $\beta_{\text{step}} > 0, \beta_{\max} > 0$
    output : Φ, β
1: Initialize $\beta^{k=0} \leftarrow \infty, k \leftarrow 1$
2: Set $\beta^k \leftarrow \beta^{k-1} - \beta_{\text{step}}$. Solve (4.13) to obtain Φ^k, M^k
    if (4.13) is infeasible :
        return Φ^{k-1}, β^{k-1}
3: Solve (4.14) to obtain D. Set $\beta^k \leftarrow \|DM^kD^{-1}\|_{\text{stab}}$
    if $\beta^k \leq \beta_{\max}$:
        return Φ^k, β^k
4: Set $k \leftarrow k + 1$ and return to step 2

In Algorithm 4.1, we alternate between minimizing over Φ and minimizing over D, and stop when no more progress can be made or when $\beta_{\max}$ is attained. No initial guess of D is needed; at iteration $k = 1$, $\beta^k = \infty$, and the $\|DMD^{-1}\|_{\text{stab}} \leq \beta$ constraint in (4.13) of step 2 is trivially satisfied. In Algorithm 4.2, we alternate between minimizing Φ and *randomizing* D. There are two main departures from Algorithm 4.1 due to the use of the randomizing D step:

1. An initial guess of D is required to generate $\beta^{k=1}$, which is then used as an input to the randomizing D step. $D = I$ is a natural choice, although we may also randomize or minimize over the initial D.

Algorithm 4.2 D-Φ iteration with randomizing D step

---

    input : $\beta_{\text{step}} > 0, \beta_{\max} > 0$
    output : Φ, β
1: Initialize $\beta^{k=0} \leftarrow \infty, k \leftarrow 1, D \leftarrow I$
2: Set $\beta^k \leftarrow \beta^{k-1} - \beta_{\text{step}}$. Solve (4.13) to obtain Φ^k, M^k
    if $k = 1$:
        Set $\beta^k \leftarrow \|DM^kD^{-1}\|_{\text{stab}}$
    if (4.13) is infeasible :
        Solve (4.17)
        if (4.17) is infeasible :
            return Φ^{k-1}, β^{k-1}
        else :
            Solve (4.13) to obtain Φ^k, M^k
    if $\beta^k \leq \beta_{\max}$:
        return Φ^k, β^k
3: Solve (4.17) to obtain D
4: Set $k \leftarrow k + 1$ and return to step 2

---

2. In step 2, when we cannot find a new Φ to make progress on β, instead of stopping, attempt to find a new D to make progress on β. If such a D can be found, find the new optimal Φ, then continue iterating.

Parameter β_{step} appears in both algorithms, and indicates the minimal robust stability margin improvement per step. For both algorithms, computational complexity is dominated by the Φ step problem (4.13) and D step problem (4.14) or (4.17). All of these problems can be distributedly computed, and all enjoy complexity that scales independently of global system size; thus, the complexity of the overall algorithm also scales independently of global system size.

4.5   Simulations

We use a ring of 10 subsystems with with spectral radius 3. Each subsystem is dynamically coupled to its neighbor on the ring, as follows

$$x_i(t + 1) = (A)_{i,i-1}x_{i-1}(t) + (A)_{i,i+1}x_{i+1}(t) + u_i(t)$$
$$\text{for} \quad i = 2\ldots N - 1,$$
$$x_1(t + 1) = (A)_{1,N}x_N(t) + (A)_{1,2}x_2(t) + u_1(t),$$
$$x_N(t + 1) = (A)_{N,N-1}x_{N-1}(t) + (A)_{N,1}x_1(t) + u_N(t)$$

$$(4.19)$$

Nonzero elements of system matrix A are randomly drawn. We focus on the state feedback case, with an LQR nominal performance objective with state penalty $Q_x = I$ and control penalty $Q_u = 50I$. The regulated output is $z_i = x_i + u_i$.

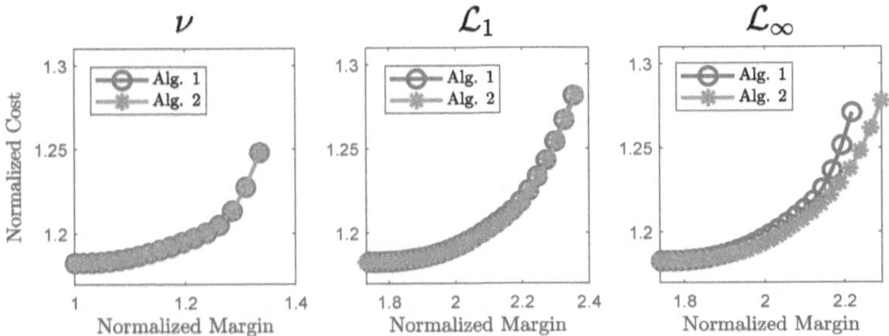

Figure 4.2: D-Φ iteration results for ν, $\mathcal{L}_1$, and $\mathcal{L}_\infty$ robust stability. Algorithm 4.1 (labelled "Alg. 1") and Algorithm 4.2 (labelled "Alg. 2") perform similarly. The controller with maximum stability margin for ν is found in 18 iterations for both algorithms; for $\mathcal{L}_1$, in 30 iterations for Algorithm 4.1 and 35 iterations for Algorithm 4.2; for $\mathcal{L}_\infty$, for $\mathcal{L}_1$, in 25 iterations for Algorithm 4.1 and 35 iterations for Algorithm 4.2.

We constrain Φ to be FIR with horizon size $T = 30$. We also constrain Φ to be sparse, such that each subsystem is only allowed to communicate to neighbors and neighbors of neighbors. We run Algorithm 4.1 and 4.2 with $\beta_{\text{step}} = 0.05$ and varying values of $\beta_{\max}$, and compare the results to the optimal LQR solution, which is unconstrained by local communication. Results are shown in Figure 4.2, where cost and robust stability margin are normalized against the LQR solution.

Both algorithms start with the standard SLS solution; in each plot, the point on the bottom left corresponds to this solution. The performance suboptimality of the SLS solution arises purely from communication constraints. There appears to be a tradeoff between nominal cost and stability margin; as the margin increases, so does the cost. Note that as the number of iterations increase, robust stability margin $\frac{1}{\beta}$ can only increase; thus, the number of iterations required increases as the desired margin $\beta_{\max}^{-1}$ increases. Algorithm 4.1 and Algorithm 4.2 perform similarly, though Algorithm 4.2 appears to require more iterations. Overall, these simulations show that the two proposed algorithms for D-Φ iteration are viable and give good margins and nominal performance, as desired. Simulation results may be reproduced using the relevant scripts in the SLS-MATLAB toolbox [6].

4.6 Conclusions and Future Work

This chapter provides a novel algorithm for distributed structured robust control. Several directions of future work may be explored:

1. This chapter focuses on DNLTV uncertainties $\mathbf{\Delta}$. The development of a similar algorithm for block-diagonal uncertainties and additional classes of uncertainty is a logical next step.

2. This chapter focuses on robust stability, but may have applications for robust performance as well. This would require a change in the optimization objective in the Φ step (and possibly the D step as well).

3. D-Φ iterations operate over a highly non-convex space; additional analysis is required to understand how the problem of local minima may be avoided, while ensuring appropriate progress. Currently, progress is enforced by decreasing β between iterations – we conjecture that this may make the algorithm more prone to local minima, particularly when the minimizing D step is employed. Alternative methods of ensuring progress should be explored.

Chapter 5

EFFICIENT DISTRIBUTED MODEL PREDICTIVE CONTROL

[1]   J. S. Li and C. Amo Alonso, "Global Performance Guarantees for Localized Model Predictive Control," *Submitted to IEEE Open Journal of Control Systems*, 2023. [Online]. Available: https://arxiv.org/abs/2303.11264,

[2]   J. S. Li, C. Amo Alonso, and J. C. Doyle, "Frontiers in Scalable Distributed Control: SLS, MPC, and beyond," in *IEEE American Control Conference*, 2021, pp. 2720–2725. DOI: 10.23919/ACC50511.2021.9483130. [Online]. Available: https://arxiv.org/abs/2010.01292,

Overview: Recent advances in model predictive control (MPC) leverage local communication constraints to produce localized MPC algorithms whose complexities scale independently of total network size. However, no characterization is available regarding global performance, i.e. whether localized MPC (with communication constraints) performs just as well as global MPC (no communication constraints). In this chapter, we provide analysis and guarantees on global performance of localized MPC — in particular, we derive sufficient conditions for optimal global performance in the presence of local communication constraints. We also present an algorithm to determine the communication structure for a given system that will preserve performance while minimizing computational complexity. The effectiveness of the algorithm is verified in simulations, and additional relationships between network properties and performance-preserving communication constraints are characterized. A striking finding is that in a network of 121 coupled pendula, each subsystem only needs to communicate with its immediate neighbors to preserve optimal global performance. Overall, this chapter offers theoretical understanding on the effect of local communication on global performance, and provides practitioners with the tools necessary to deploy localized model predictive control by establishing a rigorous method of selecting local communication constraints. This chapter also demonstrates — surprisingly — that the inclusion of severe communication constraints need not compromise global performance. Additionally, we provide an example of how we can improve the efficiency of distributed MPC by making use of a layered online-offline architecture.

5.1 Introduction

Distributed control is crucial for the operation of large-scale networks such as power grids and intelligent transport systems. Distributed model predictive control (MPC) is of particular interest, since MPC is one of the most powerful and commonly used control methods. Most formulations for distributed MPC involve open-loop policies (i.e. optimize directly over states and inputs) [16]–[23], which are computationally efficient but lack intrinsic robustness [24]. Distributed closed-loop approaches (i.e. optimize over policies) are rarer — and though they enjoy intrinsic robustness, they typically require strong assumptions, such as the existence of a static structured stabilizing controller [25] or decoupled subsystems [26]. This motivated the development of a distributed closed-loop formulation: distributed and localized MPC (DLMPC), introduced in previous work [15], [27], [28]. DLMPC is unique among distributed MPC methods in that it computes structured closed-loop policies, can be solved at scale via distributed optimization, and requires no strong assumptions on the system: it may be used on arbitrary linear systems. In the context of system level synthesis, upon which DLMPC is based, DLMPC is also the first work to extend to system level formulation to the distributed online (i.e. predictive) setting. Additionally, DLMPC enjoys minimally conservative feasibility and stability guarantees [28]. However, DLMPC requires the inclusion of local communication constraints, whose effects on performance is, thus far, unexplored — this is the focus of this chapter.

The key benefits of DLMPC are facilitated by the inclusion of local communication constraints, which are typically encapsulated in a single parameter d (rigorously defined in the next section). The question of how to select this parameter remains unresolved, as two opposing forces come into play: smaller values of d represent stricter communication constraints, which correspond to decreased complexity — however, overly strict communication constraints may render the problem infeasible, or compromise system performance. In this chapter, we address this problem by providing a rigorous characterization of the impact of local communication constraints on performance.

The analysis in this chapter is enabled by the unique formulation of DLMPC with regards to communication constraints. While previous distributed MPC approaches typically 1) focus on iterative distributed solutions of the centralized problem [16], [18], [20], [23] or 2) reshape performance objectives for distributed optimization [17], [19], DLMPC does neither — it introduces local communication constraints via the addition of a single constraint to the centralized problem, and does not require changes to the centralized performance objective. The resulting *localized MPC* problem can be exactly and optimally solved via distributed optimization. In this chapter, we rigorously analyze the effect of this local constraint by comparing the performance of the localized MPC problem to the standard global MPC problem. We restrict analysis to the linear setting, and focus on cases in which optimal global performance (with respect to any convex objective function with its minimum at the origin) may be obtained with local communication. In other words, the performance of the system is unchanged by the introduction of communication constraints. A striking finding is that in a network of coupled pendula, optimal global performance can be achieved with relatively strict local communication constraints — in fact, if every subsystem is actuated, then each subsystem only needs to communicate with its immediate neighbors to preserve optimal global performance.

For large networked systems, several studies have been conducted on the use of offline controllers with local communication constraints. Local communication can facilitate faster computational speed [29] and convergence [30], particularly in the presence of delays [31] — however, this typically comes at the cost of supoptimal global performance [32]. In [33], a trade-off between performance and decentralization level (i.e. amount of global communication) is found for a truncated linear quadratic regulator. In system level synthesis, the offline predecessor of DLMPC [3], localization is typically associated with reduced performance of around 10% relative to the global controller. More generally, for both global and localized control, the topology of the network and actuator placement [34] plays a role in achievable controller performance [35], [36] and convergence [37]. In the realm of predictive control, communication constraints are important considerations [38]. However, the improved computational speeds offered by local predictive controllers typically come at the cost of suboptimal global performance and lack of stability and convergence guarantees [39]. The novel DLMPC method [15], [27], [28] overcomes some of these drawbacks, providing both stability and convergence guarantees — however, thus far, its performance has not been substantially compared to that of the global, full-communication controller.[1] In the few instances that it has, it performed nearly identically to the global controller despite the inclusion of strict communication constraints [40], prompting further investigation.

This chapter contains two key contributions. First, we provide a rigorous charac-

---

[1]Prior work focuses on comparisons between centralized and distributed optimization schemes for localized MPC

terization of how local communication constraints restrict (or preserve) the set of trajectories available under predictive control, and use this to provide guarantees on optimal global performance for localized MPC. Secondly, we provide an exact method for selecting an appropriate locality parameter d for localized MPC. To the best of our knowledge, these are the first results of this kind on local communication constraints; our findings are useful to theoreticians and practitioners alike. A third and more minor contribution consists of a layered online-offline architecture that facilitates efficient MPC.

5.2 Localized MPC

We begin with a brief summary of global MPC and localized MPC [15], [27], [28].

Consider the standard linear time-invariant discrete-time system described in (2.1), sans disturbance $w(t)$. This system can be interpreted as N interconnected subsystems, each equipped with its own sub-controller. We model the interconnection topology as an unweighted undirected graph $\mathcal{G}_{(A,B)}(\mathcal{E}, \mathcal{V})$, where each subsystem i is identified with a vertex $v_i \in \mathcal{V}$ and an edge $e_{ij} \in \mathcal{E}$ exists whenever $[A]_{ij} \neq 0$ or $[B]_{ij} \neq 0$. The MPC problem at each timestep τ is defined as follows

$$\min_{x_t, u_t, \gamma_t} \quad \sum_{t=0}^{T-1} f_t(x_t, u_t) + f_T(x_T) \tag{5.1a}$$

$$\text{s.t.} \quad x_0 = x(\tau), \tag{5.1b}$$

$$x_{t+1} = Ax_t + Bu_t, \tag{5.1c}$$

$$x_T \in \mathcal{X}_T, \ x_t \in \mathcal{X}_t, \ u_t \in \mathcal{U}_t, \tag{5.1d}$$

$$u_t = \gamma_t(x_{0:t}, u_{0:t-1}), \ t = 0, ..., T-1 \tag{5.1e}$$

where $f_t(\cdot, \cdot)$ and $f_T(\cdot)$ are assumed to be closed, proper, and convex, with the minimum at the origin; $\gamma_t(\cdot)$ is a measurable function of its arguments; and sets $\mathcal{X}_T$, $\mathcal{X}_t$, and $\mathcal{U}$ are assumed to be closed and convex sets containing the origin for all t. We have chosen to write the standard MPC problem in this closed-loop form (i.e. optimizing over policies γ_t) due to the increased intrinsic robustness provided by closed-loop MPC versus open-loop MPC (i.e. optimizing over inputs and states directly) [24]. Policies γ_t are time-varying and capture all possible causal combinations of state x_t and input u_t.[2]

---

[2]Readers are referred to [27] for details.

In localized MPC, we impose communication constraints such that each subsystem can only communicate with a other subsystems within its local region. We shall refer to these constraints interchangeably as *local communication constraints* or *locality constraints*. Let $\mathcal{N}(i)$ denote the set of subsystems that subsystem i can communicate with. Then, to enforce locality constraints on (5.1), we replace constraint (5.1e) with

$$[u_t]_i = \gamma_{i,t}([x_{0:t}]_{j \in \mathcal{N}(i)}, [u_{0:t-1}]_{j \in \mathcal{N}(i)}) \tag{5.2}$$

Most SLS-based works use the concept of d-local neighborhoods, replacing $\mathcal{N}(i)$ with $\mathcal{N}_d(i)$:

Definition 5.1. $\mathcal{N}_d(i)$ denotes the *d-local neighborhood of subsystem i*. Subsystem $j \in \mathcal{N}_d(i)$ if there exists a path of d or less edges between subsystems i and j in $\mathcal{G}_{(A,B)}(\mathcal{E}, \mathcal{V})$.

The inclusion of local communication constraints renders problem (5.1) difficult to solve. To ameliorate this, we can apply ideas from the system level synthesis framework [3] to reparametrize the problem. We will now describe the key ideas from this framework — these ideas were already stated in II, but in the interest of clarity, we will now restate them with in an explicitly finite-horizon fashion.

First, we define $\hat{A} := \mathrm{blkdiag}(A, ..., A)$, $\hat{B} := \mathrm{blkdiag}(B, ..., B)$, and Z, the block-downshift matrix with identity matrices along the first block sub-diagonal and zeros elsewhere. Let boldface vectors and matrices be block concatenations of time-domain values, i.e.

$$\mathbf{x} = \begin{bmatrix} x_0 \\ x_1 \\ \vdots \\ x_T \end{bmatrix}, \quad \mathbf{K} = \begin{bmatrix} K_{0,0} & & & \\ K_{1,1} & K_{1,0} & & \\ \vdots & \ddots & \ddots & \\ K_{T,T} & \cdots & K_{T,1} & K_{T,0} \end{bmatrix} \tag{5.3}$$

where each $K_{i,j}$ is a matrix, and $\mathbf{K}$ denotes a causal (i.e. lower block triangular) finite horizon operators.

The behavior of system (2.1) over time horizon $t = 0 \ldots T$ can be written in terms of closed-loop maps in signal domain as

$$\mathbf{x} = (I - Z(\hat{A} + \hat{B}\mathbf{K}))^{-1}\mathbf{w} =: \mathbf{\Phi}_x \mathbf{w} \tag{5.4a}$$

$$\mathbf{u} = \mathbf{K}\mathbf{\Phi}_x \mathbf{w} =: \mathbf{\Phi}_u \mathbf{w} \tag{5.4b}$$

Note the similarity to (2.3); in our case, the formulation looks slightly different because we are explicitly making use of finite horizon operators. Theorem 2.1 may be partially restated as follows for the finite horizon case:

Proposition 5.1. For system (2.1) with causal state feedback control policy $\mathbf{u} = \mathbf{Kx}$, define $Z_{AB} := \begin{bmatrix} I - Z\hat{A} & -Z\hat{B} \end{bmatrix}$. The affine subspace of causal closed-loop maps

$$Z_{AB} \begin{bmatrix} \mathbf{\Phi}_x \\ \mathbf{\Phi}_u \end{bmatrix} = I \tag{5.5}$$

parametrizes all achievable closed-loop maps $\mathbf{\Phi}_x$, $\mathbf{\Phi}_u$.

This result allows us to reformulate a control problem over state and input signals into an equivalent problem over closed-loop maps $\mathbf{\Phi}_x$, $\mathbf{\Phi}_u$. In the case of problem (5.1), where no driving noise is present, we notice that $\mathbf{w} := \begin{bmatrix} x_0^\top & 0 & \dots & 0 \end{bmatrix}^\top$. Then, only the first block column of $\mathbf{\Phi}_x$, $\mathbf{\Phi}_u$ need to be computed. We can rewrite (5.4) as $\mathbf{x} = \mathbf{\Psi}_x x_0$ and $\mathbf{u} = \mathbf{\Psi}_u x_0$, where $\mathbf{\Psi}_x$ and $\mathbf{\Psi}_u$ correspond to the first block columns of $\mathbf{\Phi}_x$ and $\mathbf{\Phi}_u$, respectively. Then, we can rewrite (5.5) as

$$Z_{AB} \begin{bmatrix} \mathbf{\Psi}_x \\ \mathbf{\Psi}_u \end{bmatrix} = \begin{bmatrix} I \\ 0 \end{bmatrix} \tag{5.6}$$

Notice that for any $\mathbf{\Psi}_x$ satisfying constraint (5.6), $(\mathbf{\Psi}_x)_{1:N_x,:} = I$. this is due to the structure of Z_{AB}. Also, constraint (5.6) is always feasible, with solution space of dimension $N_u T$. To see this, notice that $\mathrm{rank}(Z_{AB}) = \mathrm{rank}\begin{bmatrix} Z_{AB} & \begin{bmatrix} I \\ 0 \end{bmatrix} \end{bmatrix}$ always holds, since Z_{AB} has full row rank due to the identity blocks on its diagonal; apply the Rouché-Capelli theorem [5] to get the desired result.

We now apply this closed-loop parametrization to (5.1), as is done in [27]

$$\min_{\mathbf{\Psi}_x, \mathbf{\Psi}_u} \quad f(\mathbf{\Psi}_x x_0, \mathbf{\Psi}_u x_0) \tag{5.7a}$$

$$\text{s.t.} \quad x_0 = x(\tau), \tag{5.7b}$$

$$Z_{AB} \begin{bmatrix} \mathbf{\Psi}_x \\ \mathbf{\Psi}_u \end{bmatrix} = \begin{bmatrix} I \\ 0 \end{bmatrix}, \tag{5.7c}$$

$$\mathbf{\Psi}_x x_0 \in \mathcal{X}, \quad \mathbf{\Psi}_u x_0 \in \mathcal{U} \tag{5.7d}$$

Objective f is defined such that (5.1a) and (5.7a) are equivalent; similarly, constraint sets $\mathcal{X}$, $\mathcal{U}$ are defined such that (5.1d) and (5.7d) are equivalent. Overall, problems (5.1) and (5.7) are equivalent.

In (5.7), $\mathbf{\Psi}_x$ and $\mathbf{\Psi}_u$ not only represent the closed-loop maps of the system, but also the communication structure of the system. For instance, if $[\mathbf{\Psi}_x]_{i,j} = 0$ and $[\mathbf{\Psi}_u]_{i,j} = 0$ $\forall i \neq j$, then subsystem i requires no knowledge of $(x_0)_{j \neq i}$ — consequently, no communication is required between subsystems i and j for all $j \neq i$. The relationship between closed-loop maps and communication constraints are further detailed in [27]. Thus, to incorporate local communication into this formulation, we introduce an additional constraint

$$\mathbf{\Psi}_x \in \mathcal{L}_x, \ \mathbf{\Psi}_u \in \mathcal{L}_u \tag{5.8}$$

where $\mathcal{L}_x$ and $\mathcal{L}_u$ are sets with some prescribed sparsity pattern that is compatible with the desired local communication constraints.

For simplicity, we define decision variable $\mathbf{\Psi} := \begin{bmatrix} \mathbf{\Psi}_x \\ \mathbf{\Psi}_u \end{bmatrix}$, which has $N_\Phi := N_x(T + 1) + N_u T$ rows. We also rewrite locality constraints (5.8) as $\mathbf{\Psi} \in \mathcal{L}$.

From here on, we shall use *global MPC* to refer to (5.1), or equivalently, (5.7). We shall use *localized MPC* to refer to (5.7) with constraint (5.8). We remark that localized MPC is less computationally complex than global MPC — also, for appropriately chosen locality constraints, it confers substantial scalability benefits [27].

5.3   Global Performance of Localized MPC

In this section, we analyze the effect of locality constraints $\mathbf{\Psi} \in \mathcal{L}$ on MPC performance. We are especially interested in scenarios where localized MPC achieves *optimal global performance*, i.e. $f^* = f_{\mathcal{L}}^*$, where f^* and $f_{\mathcal{L}}^*$ are the solutions to the global MPC problem and localized MPC problem, respectively, for some state x_0.

First, we must analyze the space of available trajectories from state x_0 for both global and localized MPC. We denote an available trajectory $\mathbf{y} := \begin{bmatrix} \mathbf{x}_{1:T} \\ \mathbf{u} \end{bmatrix}$.

Definition 5.2. *Trajectory set* $\mathcal{Y}(x_0)$ denotes the set of available trajectories from state x_0 under dynamics (2.1) as

$$\mathcal{Y}(x_0) := \{ \mathbf{y} : \exists \mathbf{\Psi} \text{ s.t. } Z_{AB}\mathbf{\Psi} = \begin{bmatrix} I \\ 0 \end{bmatrix}, \mathbf{y} = (\mathbf{\Psi})_{N_x+1:,:} x_0 \}$$

Localized trajectory set $\mathcal{Y}_{\mathcal{L}}(x_0)$ denotes the set of available trajectories from state x_0 under dynamics (2.1) and locality constraint $\Psi \in \mathcal{L}$ as

$$\mathcal{Y}_{\mathcal{L}}(x_0) := \{\mathbf{y} : \exists \Psi \text{ s.t. } Z_{AB}\Psi = \begin{bmatrix} I \\ 0 \end{bmatrix},$$

$$\Psi \in \mathcal{L}, \quad \mathbf{y} = (\Psi)_{N_x+1:,:}x_0\}$$

Proposition 5.2. For state x_0, if the local communication constraint set $\mathcal{L}$ is chosen such that $\mathcal{Y}(x_0) = \mathcal{Y}_{\mathcal{L}}(x_0)$, then the localized MPC problem will attain optimal global performance.

Proof. Global MPC problem (5.1) can be written as

$$\min_{\mathbf{x},\mathbf{u}} \quad f(\mathbf{x},\mathbf{u}) \tag{5.9a}$$

$$\text{s.t.} \quad x_0 = x(\tau), \ \mathbf{x} \in \mathcal{X}, \ \mathbf{u} \in \mathcal{U}, \tag{5.9b}$$

$$\mathbf{y} := \begin{bmatrix} \mathbf{x}_{1:T} \\ \mathbf{u} \end{bmatrix} \in \mathcal{Y}(x_0) \tag{5.9c}$$

The localized MPC problem can also be written in this form, by replacing $\mathcal{Y}(x_0)$ in constraint (5.9c) with $\mathcal{Y}_{\mathcal{L}}(x_0)$. Thus, if $\mathcal{Y}(x_0) = \mathcal{Y}_{\mathcal{L}}(x_0)$, the two problems are equivalent and will have the same optimal values. $\qquad\square$

We remark that this is a sufficient but not necessary condition for optimal global performance. Even if this condition is not satisfied, i.e. $\mathcal{Y}_{\mathcal{L}}(x_0) \subset \mathcal{Y}(x_0)$, the optimal global trajectory may be contained within $\mathcal{Y}_{\mathcal{L}}(x_0)$. However, this is dependent on objective f. Our analysis focuses on stricter conditions which guarantee optimal global performance for *any* objective function.

We now explore cases in which $\mathcal{Y}(x_0) = \mathcal{Y}_{\mathcal{L}}(x_0)$, i.e. the localized MPC problem attains optimal global performance. Localized trajectory set $\mathcal{Y}_{\mathcal{L}}(x_0)$ is shaped by the dynamics and locality constraints

$$Z_{AB}\Psi = \begin{bmatrix} I \\ 0 \end{bmatrix} \tag{5.10a}$$

$$\Psi \in \mathcal{L} \tag{5.10b}$$

To obtain a closed-form solution for $\mathcal{Y}_{\mathcal{L}}(x_0)$, we will parametrize these constraints. Two equivalent formulations are available. The dynamics-first formulation parametrizes constraint (5.10a), then (5.10b), and the locality-formulation parametrizes the constraints in the opposite order. The dynamics-first formulation clearly shows how local communication constraints affect the trajectory space; the locality-first formulation is less clear in this regard, but can be implemented in code with lower computational complexity than the dynamics-first formulation. We now derive each formulation.

Dynamics-first formulation

We first parametrize (5.10a), which gives a closed-form expression for trajectory set $\mathcal{Y}(x_0)$.

First, we introduce the augmented state variable. For state x_0, the corresponding augmented state X is defined as $X(x_0) := \begin{bmatrix} (x_0)_1 I & (x_0)_2 I & \dots & (x_0)_{N_x} I \end{bmatrix}$. For notational simplicity, we write X instead of $X(x_0)$; dependence on x_0 is implicit. For any matrix Λ, $\Lambda x_0 = X \vec{\Lambda}$.

Lemma 5.1. The trajectory set from state x_0 is described by

$$\mathcal{Y}(x_0) = \{\mathbf{y} : \mathbf{y} = Z_p x_0 + Z_h X \lambda, \quad \lambda \in \mathbb{R}^{N_\Phi}\}$$

$$\text{where} \quad Z_p := (Z_{AB}^\dagger)_{N_x+1:,:} \begin{bmatrix} I \\ 0 \end{bmatrix}$$

$$\text{and} \quad Z_h := (I - Z_{AB}^\dagger Z_{AB})_{N_x+1:,:}$$

and the size of the trajectory set is

$$\dim(\mathcal{Y}(x_0)) = \text{rank}(Z_h X)$$

If x_0 has at least one nonzero value, then

$$\dim(\mathcal{Y}(x_0)) = N_u T$$

Proof. As previously shown, (5.10a) always has solutions. We can parametrize the space of solutions Ψ as:

$$\Psi = Z_{AB}^\dagger \begin{bmatrix} I \\ 0 \end{bmatrix} + (I - Z_{AB}^\dagger Z_{AB})\Lambda \tag{5.11}$$

where Λ is a free variable with the same dimensions as Ψ. Recall that $\Psi_{1:N_x,:} = I$ always holds; thus, we can omit the first N_x rows of (5.11). Define $\Psi_2 := (\Psi)_{N_x+1:,:}$ and consider

$$\Psi_2 = Z_p + Z_h \Lambda \tag{5.12}$$

Combining (5.12) and the definition of $\mathbf{y}$, we have

$$\mathbf{y} = \mathbf{\Psi}_2 x_0 = Z_p x_0 + Z_h \Lambda x_0 \tag{5.13}$$

Making use of augmented state X, rewrite this as

$$\mathbf{y} = \mathbf{\Psi}_2 x_0 = Z_p x_0 + Z_h X \vec{\Lambda} \tag{5.14}$$

This gives the desired expression for $\mathcal{Y}(x_0)$ and its size.

To prove the second statement, notice that if x_0 has at least one nonzero value, then $\mathrm{rank}(Z_h X) = \mathrm{rank}(Z_h)$ due to the structure of X. All that is left is to show $\mathrm{rank}(Z_h) = N_u T$. First, note that $\mathrm{rank}(Z_{AB}) = N_x(T+1)$ due to the identity blocks on the diagonal of Z_{AB}. It follows that $\mathrm{rank}(Z_{AB}^\dagger) = \mathrm{rank}(Z_{AB}^\dagger Z_{AB}) = N_x(T+1)$. Thus, $\mathrm{rank}(I - Z_{AB}^\dagger Z_{AB}) = N_\Phi - N_x(T+1) = N_u T$. Recall that Z_h is simply $I - Z_{AB}^\dagger Z_{AB}$ with the first N_x rows removed; this does not result in decreased rank, since all these rows are zero (recall that $\mathbf{\Psi}_{1:N_x,:}$ is always equal to I). Thus, $\mathrm{rank}(Z_h) = \mathrm{rank}(I - Z_{AB}^\dagger Z_{AB}) = N_u T$. $\qquad\square$

To write the closed form of localized trajectory set $\mathcal{Y}_{\mathcal{L}}(x_0)$, we require some definitions:

Definition 5.3. *Constrained vector indices* $\mathfrak{L}$ denote the set of indices of $\mathbf{\Psi}_2 := (\mathbf{\Psi})_{N_x+1:,:}$ that are constrained to be zero by the locality constraint (5.10b), i.e.

$$(\vec{\mathbf{\Psi}}_2)_{\mathfrak{L}} = 0 \Leftrightarrow \mathbf{\Psi} \in \mathcal{L}$$

Let $N_{\mathfrak{L}}$ be the cardinality of $\mathfrak{L}$.

We now parametrize (5.10b) and combine this with Lemma 5.1, which gives a closed-form expression for localized trajectory set $\mathcal{Y}_{\mathcal{L}}(x_0)$.

Lemma 5.2. Assume there exists some $\mathbf{\Psi}$ that satisfies constraints (5.10a) and (5.10b). Then, the localized trajectory set from state x_0 is described by

$$\mathcal{Y}_{\mathcal{L}}(x_0) = \{\mathbf{y} : \quad \mathbf{y} = Z_p x_0 + Z_h X F^\dagger g \quad + \\ Z_h X (I - F^\dagger F)\mu, \quad \mu \in \mathbb{R}^{N_{\mathfrak{L}}}\}$$

$$\text{where} \quad F := (Z_h^{\mathrm{blk}})_{\mathfrak{L},:} \quad \text{and} \quad g := -(\vec{Z_p})_{\mathfrak{L}}$$

and the size of the localized trajectory set is

$$\dim(\mathcal{Y}_{\mathcal{L}}(x_0)) = \mathrm{rank}(Z_h X (I - F^\dagger F))$$

Proof. Using the augmented matrix of Z_h, we can write the vectorization of (5.12) as

$$\vec{\Psi}_2 = \vec{Z_p} + Z_h^{\text{blk}} \vec{\Lambda} \tag{5.15}$$

where $\vec{\Lambda}$ is a free variable. Incorporate locality constraint (5.10b) using the constrained vector indices

$$(\vec{Z_p} + Z_h^{\text{blk}} \vec{\Lambda})_{\mathcal{Q}} = 0 \tag{5.16}$$

This is equivalent to $(\vec{Z_p})_{\mathcal{Q}} + (Z_h^{\text{blk}})_{\mathcal{Q},:} \vec{\Lambda} = 0$, or $F \vec{\Lambda} = g$. We can parametrize this constraint as

$$\vec{\Lambda} = F^\dagger g + (I - F^\dagger F)\mu \tag{5.17}$$

where μ is a free variable. Plugging this into (5.14) gives the desired expression for $\mathcal{Y}_{\mathcal{L}}(x_0)$ and its size. We remark that there is no need to consider $\Psi_1 = I$ in relation to the locality constraints, since the diagonal sparsity pattern of the identity matrix (which corresponds to self-communication, i.e. subsystem i "communicating" to itself) satisfies any local communication constraint. $\square$

Theorem 5.1. If x_0 has at least one nonzero value, then localized MPC attains optimal global performance if

1. There exists some Ψ that satisfies constraints (5.10a) and (5.10b), and

2. $\text{rank}(Z_h X(I - F^\dagger F)) = N_u T$

Proof. By definition, $\mathcal{Y}_{\mathcal{L}}(x_0) \subseteq \mathcal{Y}(x_0)$. Equality is achieved if and only if the two sets are of equal size. Applying Lemmas 5.1 and 5.2 shows that the conditions of the theorem are necessary and sufficient for $\mathcal{Y}_{\mathcal{L}}(x_0)$ and $\mathcal{Y}(x_0)$ to be equal. Then, apply Proposition 5.2 for the desired result. $\square$

This theorem provides a criterion to assess how the inclusion of locality constraints affects the trajectory set. It also allows us to gain some intuition on the effect of these constraints, which are represented by the matrix F. If no locality constraints are included, then F has rank 0; in this case, $\text{rank}(Z_h X(I - F^\dagger F)) = \text{rank}(Z_h X) = N_u T$, via Lemma 5.1. The rank of F increases as the number of locality constraints increases; this results in decreased rank for $I - F^\dagger F$, and possibly also decreased rank for $Z_h X(I - F^\dagger F)$. However, due to the repetitive structure of $Z_h X$, this is not always the case: it is possible to have locality constraints that do not lead to decreased rank for $Z_h X(I - F^\dagger F)$. We provide a detailed numerical example of this later in this section.

Unfortunately, checking the conditions of Theorem 5.1 is computationally expensive. In particular, we must assemble and compute the rank of matrix $Z_h X (I - F^\dagger F)$. The complexity of this operation is dependent on the size of the matrix, which increases as Ψ becomes more sparse (as enforced by locality constraints). This is a problem, since it is generally preferable to use very sparse Ψ, as previously mentioned — this would correspond to an extremely large matrix $Z_h X (I - F^\dagger F)$, which is time-consuming to compute with. Ideally, sparser Ψ should instead correspond to *lower* complexity; this is the motivation for the next formulation.

Locality-first formulation

We first parametrize (5.10b), then (5.10a). This directly gives a closed-form expression for localized trajectory set $\mathcal{Y}_\mathcal{L}(x_0)$. First, some definitions:

Definition 5.4. *Support vector indices* $\mathfrak{M}$ denote the set of indices of $\vec{\Phi}$ such that $(\vec{\Phi})_\mathfrak{M} \neq 0$ is compatible with locality constraint (5.10b). Let $N_\mathfrak{M}$ be the cardinality of $\mathfrak{M}$.

Notice that this is complementary to Definition 5.3. Instead of looking at which indices are constrained to be zero, we now look at which indices are allowed to be nonzero. A subtlety is that this definition considers the entirety of Ψ, while Definition 5.3 omits the first N_x rows of Ψ.

Lemma 5.3. Assume there exists some Ψ that satisfies constraints (5.10a) and (5.10b). Then, the localized trajectory set from state x_0 is described by

$$\mathcal{Y}_\mathcal{L}(x_0) = \{ \mathbf{y} : \quad \mathbf{y} = (X_2)_{:,\mathfrak{M}} H^\dagger k \quad + \\ (X_2)_{:,\mathfrak{M}}(I - H^\dagger H)\gamma, \quad \gamma \in \mathbb{R}^{N_\mathfrak{M}} \}$$

$$\text{where} \quad X_2 := (X)_{N_x+1:,:}$$

$$\text{and} \quad H := (Z_{AB}^{\text{blk}})_{:,\mathfrak{M}} \quad \text{and} \quad k := \begin{bmatrix} \vec{I} \\ 0 \end{bmatrix}$$

and the size of the localized trajectory set is

$$\dim(\mathcal{Y}_\mathcal{L}(x_0)) = \text{rank}((X_2)_{:,\mathfrak{M}}(I - H^\dagger H))$$

Proof. Future trajectory $\mathbf{y}$ can be written as

$$\mathbf{y} = X_2 \vec{\Phi} = (X_2)_{:,\mathfrak{M}}(\vec{\Phi})_\mathfrak{M} \tag{5.18}$$

where the first equality arises from the definitions of $\mathbf{y}$ and X, and the second equality arises from the fact that zeros in $\vec{\Phi}$ do not contribute to $\mathbf{y}$; thus, we only need to consider nonzero values $(\vec{\Phi})_{\mathfrak{M}}$.

Using the augmented matrix of Z_{AB}, constraint (5.10a) can be rewritten as $Z_{AB}^{\text{blk}}\,\vec{\Phi} = k$. Nonzero values $(\vec{\Phi})_{\mathfrak{M}}$ must obey

$$H(\vec{\Phi})_{\mathfrak{M}} = k \tag{5.19}$$

Constraint (5.19) is feasible exactly when constraints (5.10a) and (5.10b) are feasible. By assumption, solutions exist, so we can parametrize the solution space as

$$(\vec{\Phi})_{\mathfrak{M}} = H^{\dagger}k + (I - H^{\dagger}H)\gamma \tag{5.20}$$

where γ is a free variable. Substituting (5.20) into (5.18) gives the desired expression for $\mathcal{Y}_{\mathcal{L}}(x_0)$ and its size. $\qquad\square$

Theorem 5.2. If x_0 has at least one nonzero value, then localized MPC attains optimal global performance if

1. there exists some Ψ that satisfies constraints (5.10a) and (5.10b), and

2. $\operatorname{rank}((X_2)_{:,\mathfrak{M}}(I - H^{\dagger}H)) = N_u T$

Proof. Similar to Theorem 5.1; instead of applying Lemma 5.2, apply Lemma 5.3. $\qquad\square$

To check the conditions of Theorem 5.2, we must assemble and compute the rank of matrix $(X_2)_{:,\mathfrak{M}}(I - H^{\dagger}H)$. The complexity of this operation is dependent on the size of this matrix, which decreases as Ψ becomes more sparse (as enforced by locality constraints). This is beneficial, since it is preferable to use very sparse Ψ, which corresponds to a small matrix $(X_2)_{:,\mathfrak{M}}(I - H^{\dagger}H)$ that is easy to compute with. This is in contrast with the previous formulation, in which sparser Ψ corresponded to increased complexity. However, from a theoretical standpoint, this formulation does not provide any intuition on the relationship between the trajectory set $\mathcal{Y}(x_0)$ and the localized trajectory set $\mathcal{Y}_{\mathcal{L}}(x_0)$.

For completeness, we now use the locality-first formulation to provide a closed-form expression of $\mathcal{Y}(x_0)$. The resulting expression is equivalent to — though decidedly more convoluted than — the expression in Lemma 5.1:

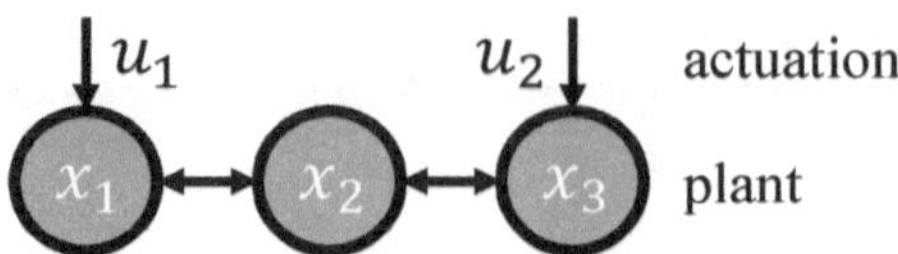

Figure 5.1: Example system with three dynamically coupled subsystems, two of which are actuated.

Lemma 5.4. The trajectory set from state x_0 is described by

$$\mathcal{Y}(x_0) = \{ \mathbf{y} : \quad \mathbf{y} = X_2 Z_{AB}^{\text{blk}} k \quad + $$
$$X_2 (I - Z_{AB}^{\text{blk}\dagger} Z_{AB}^{\text{blk}}) \gamma, \quad \gamma \in \mathbb{R}^{N_\Phi} \}$$

Proof. In the absence of locality constraints, $\mathfrak{M}$ includes all indices of $\vec{\Phi}$ since all entries are allowed to be nonzero. Here, $N_{\mathfrak{M}} = N_\Phi$, $(X_2)_{:,\mathfrak{M}} = X_2$, $(\vec{\Phi})_{:,\mathfrak{M}} = \vec{\Phi}$, and $H = Z_{AB}^{\text{blk}}$. Substitute these into the expression in Lemma 5.3 to obtain the desired result. $\qquad\square$

Notice that by definition, $X_2 Z_{AB}^{\text{blk}} k = Z_p x_0$ and $X_2 (I - Z_{AB}^{\text{blk}\dagger} Z_{AB}^{\text{blk}}) = Z_h X$. Substituting these quantities into Lemma 5.4 recovers Lemma 5.1.

Numerical example

To provide some intuition on the results from the previous subsections, we present a simple numerical example. We work with a system of three subsystems in a chain interconnection, as shown in Figure 5.1. The system matrices are

$$A = \begin{bmatrix} 1 & 2 & 0 \\ 3 & 4 & 5 \\ 0 & 6 & 7 \end{bmatrix}, B = \begin{bmatrix} 1 & 0 \\ 0 & 0 \\ 0 & 1 \end{bmatrix} \tag{5.21}$$

We set initial state $x_0 = \begin{bmatrix} 1 & 1 & 1 \end{bmatrix}^\top$, and choose a predictive horizon size of $T = 1$. Here, $N_\Phi = N_x(T + 1) + N_u T = 8$. We choose locality constraints $\mathcal{L}$ such that each subsystem may only communicate with its immediate neighbors; subsystem 1 with subsystem 2, subsystem 2 with both subsystems 1 and 3, and subsystem 3 with subsystem 2. Then, locality constraint (5.10b) is equivalent to

$$
\Psi =
\begin{bmatrix}
* & * & 0 \\
* & * & * \\
0 & * & * \\
* & * & 0 \\
* & * & * \\
0 & * & * \\
* & * & 0 \\
0 & * & *
\end{bmatrix}
\tag{5.22}
$$

where $*$ indicate values that are allowed to be nonzero. The support vector indices are $\mathfrak{M} = \{1, 2, 4, 5, 7, 9 - 16, 18, 19, 21, 22, 24\}$, and the constrained vector indices are $\mathfrak{L} = \{3, 5, 11, 14\}$ (recall that indices in $\mathfrak{L}$ do not include the first N_x rows of Ψ). We confirm that there exists some Ψ that satisfies both dynamics constraints (5.10a) and locality constraints (5.10b) by checking that constraint (5.19) is feasible.

We start with dynamics-first formulation. In our case,

$$
Z_h = \begin{bmatrix} 0_3 & c_1 & 0 & c_2 & c_1 & c_2 \end{bmatrix}
\tag{5.23}
$$

where c_1 and c_2 are defined as

$$
c_1 := \frac{1}{2}
\begin{bmatrix} 1 \\ 0 \\ 0 \\ 1 \\ 0 \end{bmatrix}
, c_2 := \frac{1}{2}
\begin{bmatrix} 0 \\ 0 \\ 1 \\ 0 \\ 1 \end{bmatrix}
\tag{5.24}
$$

Z_h has a rank of 2. Then, $Z_h X = \begin{bmatrix} Z_h & Z_h & Z_h \end{bmatrix}$, also has a rank of 2. Per Lemma 5.1, this is the size of the trajectory set $\mathcal{Y}(x_0)$, which is exactly equal to $N_u T$, as expected. We write out $Z_h X$ and $Z_h X(I - F^\dagger F)$ in full in equations (5.25) and (5.26), where zeros represent zero blocks, and their subscripts indicate the number of columns in the zero block if the number of columns is greater than 1. Boxed columns represent columns zeroed out as a result of locality constraints, i.e. if we replace the boxed columns in $Z_h X$ with zeros, we obtain $Z_h X(I - F^\dagger F)$. In our example, $Z_h X(I - F^\dagger F)$ also has a rank of 2; by Theorem 5.1, the local trajectory set is equal to the trajectory set, and by Theorem 5.1, the localized MPC problem attains optimal global performance.

$$Z_h X = \begin{bmatrix} 0_3 & c_1 & 0 & \boxed{c_2} & c_1 & \boxed{c_2} & 0_3 & c_1 & 0 & c_2 & c_1 & c_2 & 0_3 & \boxed{c_1} & 0 & c_2 & \boxed{c_1} & c_2 \end{bmatrix} \quad (5.25)$$

$$Z_h X(I - F^\dagger F) = \begin{bmatrix} 0_3 & c_1 & 0_2 & c_1 & 0_4 & c_1 & 0 & c_2 & c_1 & c_2 & 0_5 & c_2 & 0 & c_2 \end{bmatrix} \quad (5.26)$$

$$(X_2)_{:,\mathfrak{M}}(I - H^\dagger H) = \begin{bmatrix} 0_2 & c_1 & 0 & c_1 & 0_3 & c_1 & 0 & c_2 & c_1 & c_2 & 0_3 & c_2 & c_2 \end{bmatrix} \quad (5.27)$$

Two observations are in order. First, we notice that the rank of $Z_h X$ ($= 2$) is low compared to the number of nonzero columns ($= 12$), especially when x_0 is dense. Additionally, the structure of $Z_h X$ is highly repetitive; the only two linearly independent columns are c_1 and c_2, and each appears 6 times in $Z_h X$. Furthermore, the specific values of x_0 do not affect the rank of these matrices — only the placement of nonzeros and zeros in x_0 matters.

Second, notice that post-multiplying $Z_h X$ by $(I - F^\dagger F)$ effectively zeros out columns of $Z_h X$. However, due to the repetitive structure of $Z_h X$, this does not result in decreased rank for $Z_h X(I - F^\dagger F)$. In fact, it is difficult to find a feasible locality constraint that results in decreased rank. This observation is corroborated by simulations in the later portion of this chapter, in which we find that locality constraints that are feasible also typically preserve global performance. For more complex systems and larger predictive horizons, post-multiplication of $Z_h X$ by $(I - F^\dagger F)$ no longer cleanly corresponds to zeroing out columns, but similar intuition applies.

We now apply the locality-first formulation. To check the conditions of Theorem 5.2, we must construct the matrix $(X_2)_{:,\mathfrak{M}}(I - H^\dagger H)$ and check its rank. This matrix is written out in equation (5.27). As expected, the rank of this matrix is also equal to two. Additionally, notice that $(X_2)_{:,\mathfrak{M}}(I - H^\dagger H)$ contains the same nonzero columns as $Z_h X(I - F^\dagger F)$: c_1 and c_2 are each repeated four times, in slightly different orders. This is unsurprising, as the two formulations are equivalent.

5.4 Algorithmic Implementation of Optimal Locality Selection

Leveraging the results of the previous section, we introduce an algorithm that selects the appropriate locality constraints for localized MPC. For simplicity, we restrict ourselves to locality constraints corresponding to d-local neighborhoods, though we remark that Subroutine 5.1 is applicable to arbitrary communication structures.

The localized MPC problem can be solved via distributed optimization techniques; the resulting distributed and localized MPC problem enjoys complexity that scales with locality parameter d, as opposed to network size N [27]. Thus, when possible, it is preferable to use small values of d to minimize computational complexity. For a given system and predictive horizon length, Algorithm 5.2 will return the *optimal locality size d* — the smallest value of d that attains optimal global performance.

As previously described, the specific values of x_0 do not matter — only the placement of nonzeros and zeros in x_0 matters. We will restrict ourselves to considering dense values of x_0. For simplicity, our algorithm will work with the vector of ones as x_0 — the resulting performance guarantees hold for *any* dense x_0.

To check if a given locality constraint preserves global performance, we must check the two conditions of Theorem 5.2. First, we must check whether there exists some Ψ that satisfies both dynamics and locality constraints; this is equivalent to checking whether (5.19) is feasible. We propose to check whether

$$\|H(H^\dagger k) - k\|_\infty \leq \epsilon \tag{5.28}$$

for some tolerance ϵ. Condition (5.28) can be distributedly computed due to the block-diagonal structure of H. Define partitions $[H]_i$ such that

$$H = \text{blkdiag}([H]_1, [H]_2 \dots [H]_N) \tag{5.29}$$

In general, H has N_x blocks, where N_x is the number of states. Since $N \leq N_x$ and one subsystem may contain more than one state, we are able to partition H into N blocks as well. Then, (5.28) is equivalent to

$$\|[H]_i([H]_i^\dagger [k]_i) - [k]_i\|_\infty \leq \epsilon \quad \forall i \tag{5.30}$$

If this condition is satisfied, then it remains to check the second condition of Theorem 5.2. To so, we must construct matrix $J := (X_2)_{:,\mathfrak{M}}(I - H^\dagger H)$ and check its rank. Notice that J can be partitioned into submatrices J_i, i.e. $J := \begin{bmatrix} J_1 & J_2 & \dots & J_N \end{bmatrix}$, where each block J_i can be constructed using only information from subsystem i, i.e. $[H]_i$, $[k]_i$, etc. Thus, J can be constructed in parallel — each subsystem i performs Subroutine 5.1 to construct J_i.

Subroutine 5.1 checks whether the dynamics and locality constraints are feasible by checking (5.30), and if so, returns the appropriate submatrix J_i. Notice that the quantity $[H]_i$ is used in both the feasibility check and in J_i. Also, x_0 does not appear, as we are using the vector of ones in its place.

Having obtained J corresponding to a given locality constraint, we need to check its rank to verify whether global performance is preserved, i.e. $\text{rank}(J) = N_u T$, as per

---

Subroutine 5.1 Local sub-matrix for subsystem i

---

   input : $[H]_i$, $[k]_i$, ϵ
   output : J_i
1: Compute $w = [H]_i^\dagger [k]_i$
2: **if** $\| [H]_i w - [k]_i \|_\infty > \epsilon$ **:**
    $J_i \leftarrow$ False
   else :
    $J_i \leftarrow I - [H]_i^\dagger [H]_i$
   return J_i

Theorem 5.2. As previously described, we restrict ourselves to locality constraints of the d-local neighborhood type, preferring smaller values of d as these correspond to lower complexity for the localized MPC algorithm [27]. Thus, in Algorithm 5.2, we start with the smallest possible value of $d = 1$, i.e. subsystems communicate only with their immediate neighbors. If $d = 1$ does not preserve global performance, we iteratively increment d, construct J, and check its rank, until optimal global performance is attained.

Algorithm 5.2 Optimal local region size

---

   input : A, B, T, ϵ
   output : d
   for $d = 1 \ldots N$ **:**
1:    **for** $i = 1 \ldots N$ **:**
      Construct $[H]_i$, $[k]_i$
      Run Subroutine 5.1 to obtain J_i
      if J_i is False **:**
         continue
2:    Construct $J = \begin{bmatrix} J_1 & \ldots & J_N \end{bmatrix}$
3:    **if** $\mathrm{rank}(J) = N_u T$ **:**
      return d

In step 1 of Algorithm 5.2, we call Subroutine 5.1 to check for feasibility and construct submatrices J_i. If infeasibility is encountered, or if optimal global performance is not attained, we increment d; otherwise, we return optimal locality size d.

Complexity

To analyze complexity, we first make some simplifying scaling assumptions. Assume that the number of states N_x and inputs N_u are proportional to the number of subsystems N, i.e. $O(N_x + N_u) = O(N)$. Also, assume that the number of nonzeros

$N_{\mathfrak{M}}$ for locality constraint corresponding to parameter d are related to one another as $O(N_{\mathfrak{M}}) = O(NdT)$.

Steps 1 and 3 determine the complexity of the algorithm. In step 1, each subsystem performs operations on matrix $[H]_i$, which has size of approximately $N_x(T + 1)$ by dT — the complexity will vary depending on the underlying implementations of the pseudoinverse and matrix manipulations, but will generally scale according to $O((dT)^2 N_x T)$, or $O(T^3 d^2 N)$. In practice, d and T are typically much smaller than N, and this step is extremely fast; we show this in the next section.

In step 3, we perform a rank computation on a matrix of size $(N_x + N_u)T$ by $N_{\mathfrak{M}}$. The complexity of this operation, if Gaussian elimination is used, is $O((N_x + N_u)^{1.38} T^{1.38} N_{\mathfrak{M}})$, or $O(T^{2.38} N^{2.38} d)$. Some speedups can be attained by using techniques from [41], which leverage sparsity — typically, J is quite sparse, with less than 5% of its entries being nonzero. In practice, step 3 is the dominating step in terms of complexity.

We remark that this algorithm needs only to be run once offline for any given localized MPC problem. Given a system and predictive horizon, practitioners should first determine the optimal locality size d using Algorithm 5.2, then run the appropriate online algorithm from [27].

5.5 Simulations

We first present simulations to supplement runtime characterizations of Algorithm 5.2 from the previous section. Then, we use the algorithm to investigate how optimal locality size varies depending on system size, actuation density, and prediction horizon length. We find that optimal locality size is primarily a function of actuation density. We also verify in simulation that localized MPC performs identically to global MPC when we use the optimal locality size provided by Algorithm 5.2, as expected. Code needed to replicate all simulations can be found at `https://github.com/flyingpeach/LocalizedMPCPerformance`. This code makes use of the SLS-MATLAB toolbox [6], which includes an implementation of Algorithm 5.2.

We note that when implementing subroutine 5.1 in MATLAB, use of the backslash operator (i.e. `H\k`) is faster than the standard pseudoinverse function (i.e. `pinv(H)*k`). The backslash operator also produces J_i matrices that are as sparse as possible, which facilitates faster subsequent computations.

System and parameters

We start with a two-dimensional n by n square mesh. Every pair of neighboring subsystems have a 40% probability to be connected by an edge — the expected number of edges is $0.8 \times n \times (n-1)$. We consider only fully connected graphs. Each subsystem i represents a two-state subsystem of linearized and discretized swing equations

$$\begin{bmatrix} \theta(t+1) \\ \omega(t+1) \end{bmatrix}_i = \sum_{j \in \mathcal{N}_1(i)} [A]_{ij} \begin{bmatrix} \theta(t) \\ \omega(t) \end{bmatrix}_j + [B]_i [u]_i \qquad (5.31\text{a})$$

$$[A]_{ii} = \begin{bmatrix} 1 & \Delta t \\ -\frac{k_i}{m_i} \Delta t & 1 - \frac{d_i}{m_i} \Delta t \end{bmatrix},$$

$$[A]_{ij} = \begin{bmatrix} 0 & 0 \\ \frac{k_{ij}}{m_i} \Delta t & 0 \end{bmatrix}, [B]_i = \begin{bmatrix} 0 \\ 1 \end{bmatrix} \qquad (5.31\text{b})$$

where $[\theta]_i$, $[\omega]_i$, and $[u]_i$ are the phase angle deviation, frequency deviation, and control action of the controllable load of bus i. Parameters are m_i^{-1} (inertia), d_i (damping), and k_{ij} (coupling); they are randomly drawn from uniform distributions over $[0, 2]$, $[1, 1.5]$ and $[0.5, 1]$ respectively. Self term k_i is defined as $k_i :=$ $\sum_{j \in \mathcal{N}_1(i)} k_{ij}$. Discretization step size Δt is 0.2.

Under the given parameter ranges, the system is typically neutrally stable, with a spectral radius of 1. The baseline system parameters are $n = 5$ (corresponding to a 5 grid containing 25 subsystems, or 50 states) and 100% actuation. Note that this does not correspond to "full actuation" in the standard sense; it means that each subsystem (which contains 2 states) has one actuator — 50% of states are actuated. We use a prediction horizon length of $T = 15$ unless otherwise stated.

Algorithm runtime

We plot the runtime of Algorithm 5.2 in Figure 5.2 for different system sizes and horizon lengths. We separately consider runtimes for matrix construction (step 1) and rank determination (step 3). The former is parallelized, while the latter is not.

Runtime for matrix construction is extremely small. Even for the grid with 121 subsystems (242 states), this step takes less than a millisecond. Interestingly, matrix construction runtime also stays relatively constant with increasing network size, despite the worst-case runtime scaling linearly with N, as described in the previous section. This is likely due to the sparse structure of H. Conversely, matrix construction runtime increases with increasing horizon length.

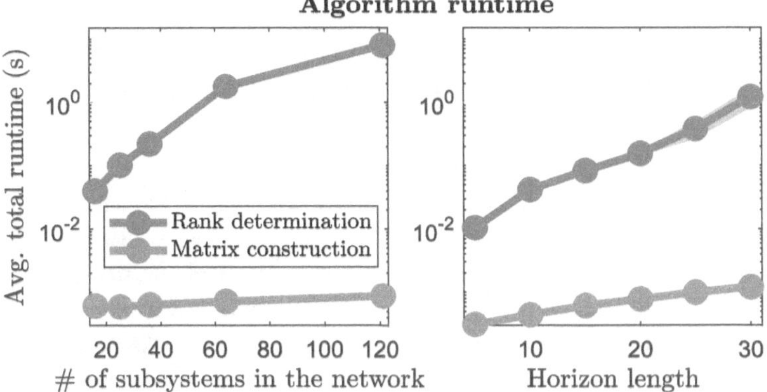

Figure 5.2: Runtime of matrix construction (step 1, green) and rank determination (step 3, pink) of Algorithm 5.2 vs. network size and horizon length. Parallelized (i.e. per-subsystem) runtimes are shown for matrix construction. The algorithm was run for grids containing 16, 25, 36, 64, and 121 subsystems. For each point, we run the algorithm on five different systems, and plot the average and standard deviation — here, the standard deviation is so small that it is barely visible. As expected, the rank determination step dominates total runtime, while the matrix construction step is extremely fast.

Runtime for rank determination dominates total algorithm runtime, and increases with both system size and horizon length. Rank determination runtime appears to increase more sharply with increasing horizon length than with increasing system size. Further runtime reductions may be achieved by taking advantage of techniques described in the previous section; however, even without additional speedups, the runtime is no more than 10 seconds for the grid with 242 states.

Optimal locality size as a function of system parameters

We characterize how optimal locality size changes as a function of the system size and horizon length — the results are summarized in Figure 5.3. Actuation density is the main factor that affects optimal locality size. Remarkably, at 100% actuation, the optimal locality size always appears to be $d = 1$, the smallest possible size (i.e. communication only occurs between subsystems that share an edge). As we decrease actuation density, the required optimal locality size increases. This makes sense, as unactuated subsystems must communicate to at least the nearest actuated subsystem, and the distance to the nearest actuated subsystem grows as actuation density decreases.

The optimal locality size also increases as a function of system size — but only when we do not have 100% actuation. At 60% actuation, for 121 subsystems, the optimal locality size is around $d = 5$; this still corresponds to much less communication than global MPC.

Predictive horizon length does not substantially impact optimal locality size. At short horizon lengths ($T \leq 10$), we see some small correlation, but otherwise, optimal locality size stays constant with horizon size. Similarly, the stability of the system (i.e. spectral radius) appears to not affect optimal locality size; this was confirmed with simulations over systems with spectral radius of 0.5, 1.0, 1.5, 2.0, and 2.5 for 60%, 80%, and 100% actuation (not included in the plots).

Localized performance

From the previous section, we found that with 100% actuation, the optimal locality size is always 1. This means that even in systems with 121 subsystems, each subsystem need only communicate with its immediate neighbors (i.e. 4 or less other subsystems) to attain optimal global performance. This is somewhat surprising, as this is a drastic (roughly 30-fold) reduction in communication compared to global MPC. To confirm this result, we ran simulations on 20 different systems of size $N = 121$ and 100% actuation. We use LQR objectives with random positive diagonal matrices Q and R, and state bounds $\theta_i \in [-4, 4]$ for phase states and $\omega_i \in [-20, 20]$ for frequency states. We use random initial conditions where each value $(x_0)_i$ is drawn from a uniform distribution over $[-2, 2]$.

For each system, we run localized MPC with $d = 1$, then global MPC, and compare their costs over a simulation of 20 timesteps. Over 20 simulations, we find the maximum cost difference between localized and global MPC to be 5.6e-6. Thus, we confirm that the reported optimal locality size is accurate, since the cost of localized and global MPC are nearly identical.

Feasibility vs. optimality of locality constraints

In Algorithm 5.2, a given locality size is determined to be suboptimal if (1) the locality constraints are infeasible, or (2) the locality constraints are feasible, but matrix J has insufficient rank. The numerical example from earlier in this chapter suggests that the second case is rare — to further investigate, we performed 200 random simulations, in which all parameters were randomly selected from uniform

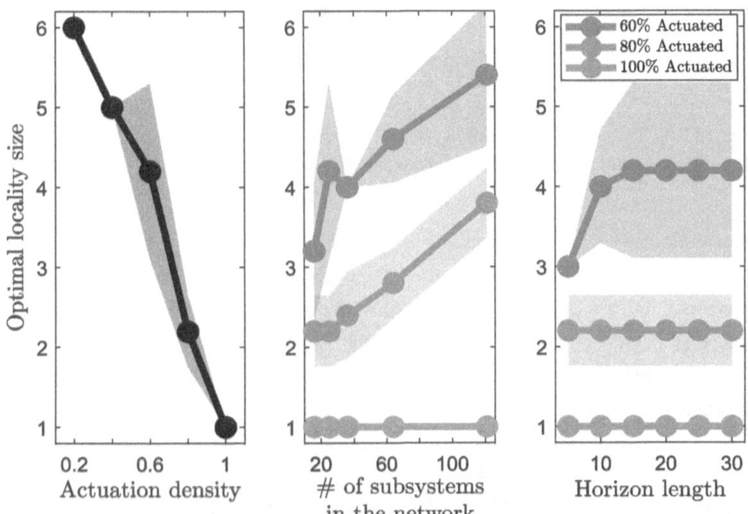

Figure 5.3: Optimal locality size as a function of various parameters. Each point represents the average over five different systems; standard deviations are shown by the fill area. **(Left)** Optimal locality size vs. actuation density. The two are inversely correlated. **(Center)** Optimal locality size vs. network size for 60% actuation (pink), 80% actuation (blue), and 100% actuation (green). For 60% and 80% actuation, optimal locality size roughly increases with network size. For 100% actuation, the optimal locality size is always 1, independent of network size. **(Right)** Optimal locality size vs. predictive horizon length for 60% actuation (pink), 80% actuation (blue), and 100% actuation (green). For 60% and 80% actuation, optimal locality size increases with horizon size up until $T = 10$, then stays constant afterward. For 100% actuation, the optimal locality size is always 1.

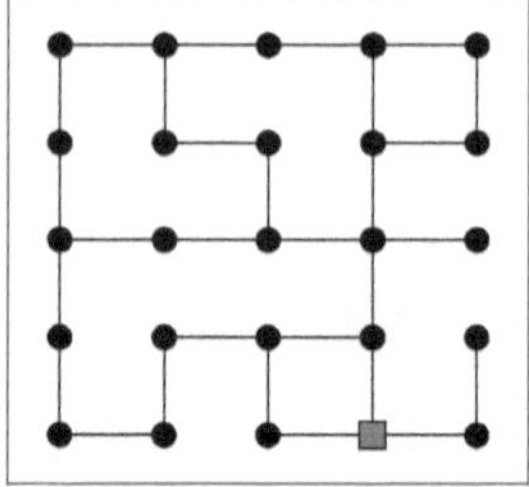

Figure 5.4: Topology of example system. We will plot the time trajectories of states, disturbances, and input for the red square subsystem.

distributions — grid size N from $[4, 11]$ (corresponding to system sizes of up to 121 subsystems), actuation density from $[0.2, 1.0]$, spectral radius from $[0.5, 2.5]$ and horizon length from $[3, 20]$. In these 200 simulations, we encountered 4 instances where a locality constraint was feasible but resulted in insufficient rank; in the vast majority of cases, if a locality constraint was feasible, the rank condition was satisfied as well.

5.6 Efficient Two-Layer MPC

In the previous portion of this chapter, we described how distributed MPC can perform just as well as centralized MPC, even under local communication constraints. However, despite the scalability of the distributed and localized MPC problem, MPC itself still incurs computational burden, as it requires an optimization problem to be solved at each timestep. To lower the required computation, we suggest a layered online-offline architecture based on existing works in the literature [42]; our two-layer controller is unique in that it can be fully synthesized and implemented in a localized manner.

We use a system of linearized swing equations embedded in a square mesh, similar to the system used in the previous section. An example topology is shown in Figure 5.4. We use dynamics from (5.31), replacing (5.31a) with

$$\begin{bmatrix} \theta(t+1) \\ \omega(t+1) \end{bmatrix}_i = \sum_{j \in \mathcal{N}_1(i)} [A]_{ij} \begin{bmatrix} \theta(t) \\ \omega(t) \end{bmatrix}_j + [B]_i \left([u]_i + [w]_i \right) \tag{5.32}$$

where the only difference is that we have added a disturbance term w. The parameters are drawn over slightly slightly different distributions: we now set damping d_i to

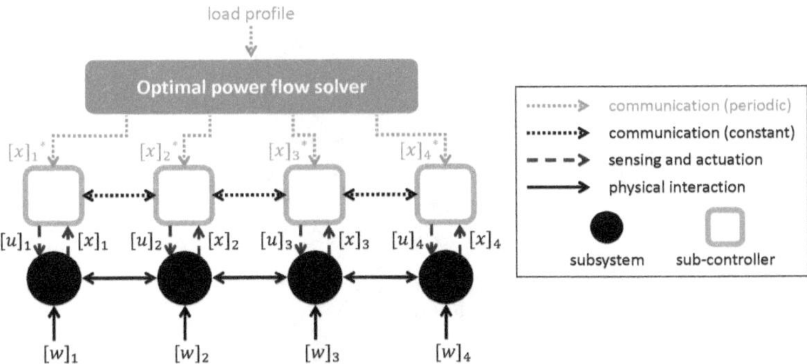

Figure 5.5: Architecture of example system with optimal power flow solver, sub-controllers, and subsystems. For ease of visualization, we depict a simple 4-subsystem topology instead of the 25-subsystem mesh we'll be using.

zero and draw inertia m_i^{-1} from a random uniform distribution over $[0, 10]$. These two changes both make this system more challenging to control compared to the system of the previous section; the decreased damping and inertia render the system unstable, and in our example the spectral radius of the A matrix is 1.5. We define the state at subsystem i:

$$[x]_i := \begin{bmatrix} \theta(t + 1) \\ \omega(t + 1) \end{bmatrix}_i \tag{5.33}$$

We will incorporate optimal power flow into the system as follows: periodically, we will generate a random load profile, and solve a direct-current optimal power flow problem to obtain the optimal state setpoint x^*. We will then send each sub-controller i its individual optimal setpoint $[x]_i^*$, and expect each subsystem to reach this setpoint. We add local communication constraints: each sub-controller may only communicate with their immediate neighbors (i.e. subsystems with whom they share an edge) and neighbors of those neighbors, i.e. the local region size d is set to 2. Controllers must reject frequent, random disturbances which have much smaller magnitudes than that of the setpoint changes. Additionally, we enforce actuator saturation constraints of the form $\|u(t)\|_\infty \leq u_{max}$. Overall, this simple system captures some common challenges from power grid control — namely, instability, setpoint changes, actuator saturation, and disturbances. A diagram of the system setup is shown in Figure 5.5.

We propose a distributed two-layer MPC scheme for this system, as shown in Figure

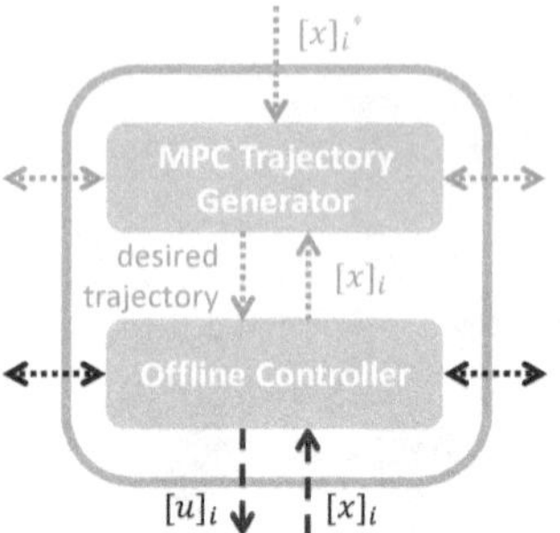

Figure 5.6: Layered sub-controller for the ith subsystem. Horizontal dotted lines indicate communication with neighboring sub-controllers.

5.6. We decompose the main problem into two subproblems: reacting to setpoint changes and rejecting disturbances. We use MPC to deal with setpoint changes — specifically, we use MPC in the top layer to generate a desired trajectory that will accomplish the indicated setpoint change while respecting actuator saturation. In the bottom layer, we use a linear offline controller to track the trajectory generated by the top layer, while rejecting disturbances.

To facilitate scalability, we use the distributed and localized MPC technique previously described [27] in the top layer, and a standard SLS controller [3] in the bottom layer. The resulting two-layer controller is fully distributed; each subsystem synthesizes and implements both layers of its own sub-controller. Synthesis of the two layers is independent of one another, although some synthesis parameters may be shared. Additionally, all cross-layer communication is local, i.e. a top layer sub-controller will never communicate with a bottom layer sub-controller from a different subsystem. Distributed implementation and synthesis allows our controller to enjoy a runtime that is independent of total system size, like its component SLS-based controllers.

This two-layer controller allows us to obtain MPC-like behavior, but at a fraction of the computational cost. For any MPC method — including distributed MPC — we must solve an optimization problem at every timestep. However, in this two-layer controller, we only need to solve an optimization problem when the setpoint changes. If the setpoint changes every T_{OPF} timesteps, then this two-layer controller offers a T_{OPF}-fold reduction in computational cost compared to a single MPC controller.

The reader may be tempted to ask whether the bottom layer is required at all: could

Table 5.1: LQR costs corresponding to Figure 5.7

Controller	LQR Cost
Centralized LQR	2.09e13
Centralized MPC	1.00
Distributed two-layer	1.02

we simply run MPC periodically without need of an offline controller? Unfortunately, the answer is no; in the case of an unstable system, disturbances will persist between MPC runs and severely compromise performance. The reverse question can also be asked: could we avoid MPC altogether and simply use an offline controller? Again, the answer is no; the linear offline controller alone can react poorly to setpoint changes and actuator saturation, as we will see in simulation.

We now present results from simulations for $T_{OPF} = 20$. We compared the performance of the distributed two-layer controller with the performance of centralized MPC and centralized LQR (i.e. offline controller). The centralized controllers are free of communication constraints while the layered controller is limited to local communication. The resulting LQR costs, normalized by the centralized MPC cost, are shown in Table 5.1. We plot trajectories from the red square subsystem from Figure 5.4 in Figure 5.7, focusing on a window of time during which only one setpoint change occurs.

In this example, the setpoint change induces unstable behavior from the centralized LQR. This is to be expected due to saturation-induced windup effects. The centralized MPC controller and distributed two-layer controller perform very similarly. From the previous results in this chapter, it should not be surprising that a reduction in communication does not substantially affect performance. However, this example additionally shows that we need not solve the MPC problem at every timestep to obtain MPC-like performance — here, the two-layer controller solve an MPC problem every 20 timesteps.

To check general behavior, we re-run the simulation 30 times with different randomly generated grids, plant parameters, disturbances, and load profiles. The resulting LQR costs, normalized by the cost of centralized MPC, are shown in Table 5.2. The centralized LQR controller loses stability in 4 out of 30 simulations. In these cases, as with the previous example, the two-layer controller maintains stability and performs highly similarly to centralized MPC. For the 26 simulations for which the

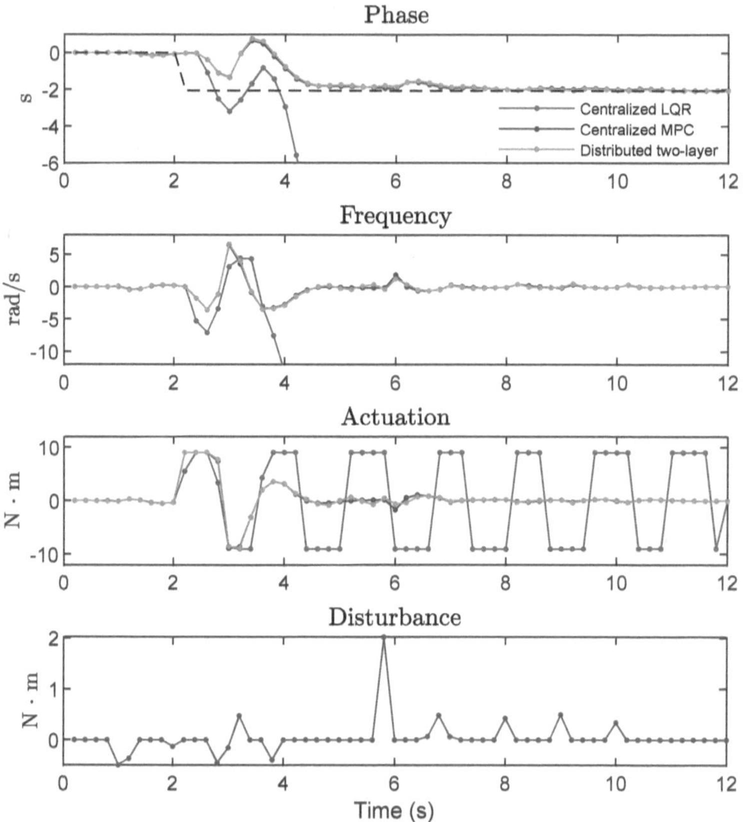

Figure 5.7: Performance of centralized LQR, centralized MPC, and distributed two-layer controller. The phase setpoint is shown by the dotted line; a setpoint change occurs at $t = 2$. The LQR controller loses stability, resulting in large oscillations in phase and frequency, which are omitted from the plot after around $t = 4$; the associated actuation engages in oscillations as well, which are shown on the plot. The MPC and two-layer controllers perform similarly.

Table 5.2: LQR costs averaged over 30 simulations

Controller	LQR cost	
	All sims	LQR controller stable
Centralized LQR	1.14e7	1.05
Centralized MPC	1.00	1.00
Distributed two-layer	1.01	1.01

LQR controller maintains stability, we show the costs in a separate column. In this case, the LQR controller performs similarly to the MPC and two-layer controllers.

Overall, these examples show that the distributed two-layer controller indeed achieves near-optimal performance, at a fraction of the communication and computational cost required by centralized MPC, which is the ideal (but computationally expensive) controller for this problem.

5.7 Conclusions and Future Work

In this chapter, we provided analysis and guarantees on locality constraints and global performance. We presented lemmas, theorems, and an algorithm to certify optimal global performance — these are the first results of their kind, to the best of our knowledge. We then leveraged these theoretical results to provide an algorithm that determines the optimal locality constraints that will expedite computation while preserving the performance — this is the first exact method to compute the optimal locality parameter d for DLMPC. In addition, we showed how further computational savings can be obtained by combining DLMPC with an offline SLS controller. The resulting two-layer controller performs similarly to centralized MPC at a fraction of the computational cost.

Several directions of future work may be explored:

1. The results in this chapter can be leveraged to investigate the relationship between network topology and optimal locality constraints, i.e. the strictest communication constraints that still preserve optimal global performance. Certain topologies may require long-distance communication between handful of subsystems; others may require no long-distance communications. A more thorough characterization will help us understand the properties of systems that are suited for localized MPC.

2. Algorithm 5.2 considers d-local communication constraints; a natural extension is to consider non-uniform local communication constraints, which are supported by the theory presented in this chapter. A key challenge of this research direction is the combinatorial nature of available local communication configurations; insights from the research direction suggested above will likely help narrow down said set of configurations.

3. Simulations suggest that feasibility of a given locality constraint overwhelmingly coincides with optimal global performance. This poses the question of whether feasibility of a given locality constraint is *sufficent* for J to be full rank under certain conditions, and what these conditions may be. Additional investigation could reveal more efficient implementations of Algorithm 5.2, as bypassing the rank checking step would save a substantial amount of computation time.

4. This chapter focuses on nominal trajectories. Additional investigation is required to characterize the impact of locality constraints on trajectories robust to disturbances. For polytopic disturbances, the space (or minima) of available values of Ξg from (10) in [27] is of interest. Due to the additional variables in the robust MPC problem, we cannot directly reuse techniques from this chapter, though similar ideas may be applicable.

5. Rather than creating a brand new technique or tuning existing algorithms for small improvements, we can cleverly combine existing algorithms for large improvements, as we did with the two-layer controller. A future direction to explore is the development of systematic techniques to integrate various control algorithms in this manner, to facilitate high-performing, efficient, and safe control.

Chapter 6

INTERNAL FEEDBACK IN PRIMATE CORTEX

[1] J. Stenberg, J. S. Li, A. A. Sarma, and J. C. Doyle, "Internal Feedback in Biological Control: Diversity, Delays, and Standard Theory," in *IEEE American Control Conference*, 2022, pp. 462–467. DOI: 10.23919/ACC53348.2022.9867794. [Online]. Available: http://arxiv.org/abs/2109.11752,

[2] J. S. Li, "Internal Feedback in Biological Control: Locality and System Level Synthesis," in *IEEE American Control Conference*, 2022, pp. 474–479. DOI: 10.23919/ACC53348.2022.9867769. [Online]. Available: http://arxiv.org/abs/2109.11757,

[3] J. S. Li, A. A. Sarma, T. J. Sejnowski, and J. C. Doyle, "Internal feedback in the cortical perception-action loop enables fast and accurate behavior," *Submitted to Proceedings of the National Academy of Sciences*, 2023. [Online]. Available: https://arxiv.org/abs/2211.05922,

Overview: Models of sensorimotor control have assumed that sensory information from the environment leads to actions, which then act back on the environment, creating a single, unidirectional perception-action loop. However, the existence of internal feedback projections — signals flowing from motor areas or late sensory processing regions back to early sensory processing regions such as primary visual and auditory areas — are ubiquitous and more numerous in sensorimotor structures than feedforward projections.[1] The function of this internal feedback is poorly understood; the purpose of this chapter is to develop and apply tools from control theory to explain the purpose of internal feedback. In particular, we argue that delayed, local, and disributed communication are the key to understanding the function of internal feedback. First, we introduce novel theory that incorporates delays into standard optimal control theory, and show that the resulting controller is reliant on internal feedback for compensation and estimation. Then, we use distributed control to further model internal feedback in a system subject to local communication constraints; the resulting models show that internal feedback is crucial for localization of function and behavior, and in systems with sparse actuation, necessarily outnumber feedforward projections. Finally, we show how internal feedback

---

[1]For the purposes of this chapter, we use "feedforward" in the neuroscientific sense to indicate projections from sensory areas toward motor areas. This is not to be confused with feedforward control.

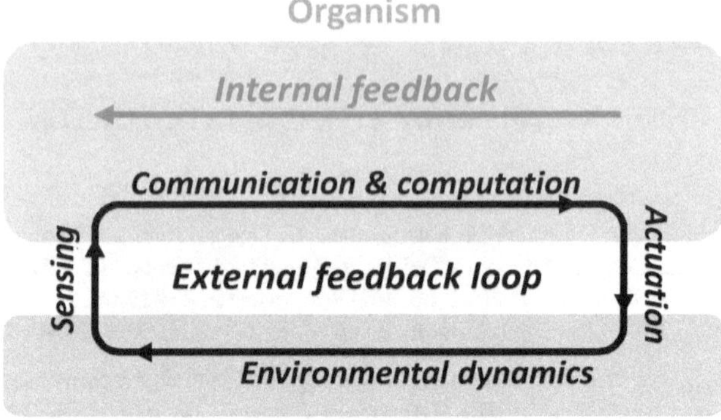

Figure 6.1: Single-loop model of sensorimotor control. The organism receives information from the external environment via sensors, communicates this information through the body, computes actions, then acts on the environment; this forms the *external feedback loop*, or single loop model (black). Internal signals that flow opposite to the direction of the external feedback loop are classified as *internal feedback* (pink). Thus, the internal feedback is *counterdirectional*. Internal feedback also includes *lateral interactions* within or between areas (not shown).

facilitates good task performance when communications and attention are subject to speed-accuracy trade-offs. Overall, our models explain anatomical, physiological and behavioral observations, including motor signals in visual cortex, heterogeneous kinetics of sensory receptors and the presence of giant Betz cells in motor cortex and Meynert cells in visual cortex of humans, as well as internal feedback patterns and unexplained heterogeneity in other neural systems.

6.1 Introduction

Feedback control is an essential strategy for both engineered and biological systems to achieve reliable movements in unpredictable environments [43]. Optimal and robust control theory, which provide a general mathematical foundation to study feedback systems, have been used successfully to explain behavioral observations by modeling the sensorimotor system as a single control loop also called the perception-action cycle or perception-action loop [44]–[46]. In these models, the sensorimotor system senses the environment, communicates signals from sensors to the brain, computes actions, and then acts on the environment, feeding back to the sensors and forming a single unidirectional loop as shown in Figure 6.1.

Consider the canonical model of localized function in the primate visuomotor cortical pathway, depicted in Figure 6.2: a visual signal is encoded on the retina, then travels to the lateral geniculate nucleus (LGN) of the thalamus, and on to the primary visual cortex (V1), progressing through successive transformations until it reaches the primary motor cortex (M1), the spinal cord, and ultimately the muscles. The single-loop feedback model also makes implicit assumptions about the interpretation of responses from sensory and motor populations of neurons, which represent sensory signals and action signals, respectively. Although intuitive, this model neglects a well-known and ubiquitous feature of sensorimotor processing: *internal feedback* [47], which is the main focus of this chapter.

The perception-action control model does not have a direct role for internal feedback connections. Internal feedback includes all signals that do not flow from sensing towards action. We can divide internal feedback into two broad categories: *counterdirectional* between brain areas and *lateral interactions* within or between areas. Counterdirectional internal feedback is in the opposite direction of the single-loop model (for instance, from V2 to V1); these signals flow from action toward sensing. Lateral internal feedback consists of recurrent connections within and between areas (for instance, from V2 to V2, or from MT to IT). This distinction emphasizes the importance of where control signals are spatially located.

The single-loop model offers a set of tools from control theory and a conceptual framework that allows subsystems to be treated as successive transformations that can be studied in isolation. However, these subsystems are not isolated. With internal feedback, each subsystem has access to both bottom-up, top-down, and lateral information. The eye is itself a site of computation and control: as the eye moves and senses different parts of the visual scene, lateral interactions within the retina control spatial and temporal filter properties that can adapt and identify important features under a wide range of illumination and scene dynamics [48], [49]. Retinal ganglion cells project to relay neurons in the LGN, which then project to primary visual cortex, V1, but a much greater number of feedback neurons project from V1 to LGN [50]–[52] (Figure 6.2).

Projections from motor areas in cortex to visual areas have a wide range of morphology, myelination and synaptic kinetics [50], [53], [54]. Given the position of M1 in the final common pathway, one might expect activity in M1 to be driven by current visual stimuli or current movements, but instead autonomous or top-down preparatory activity with internal dynamics dominate the data [55]. Counterintu-

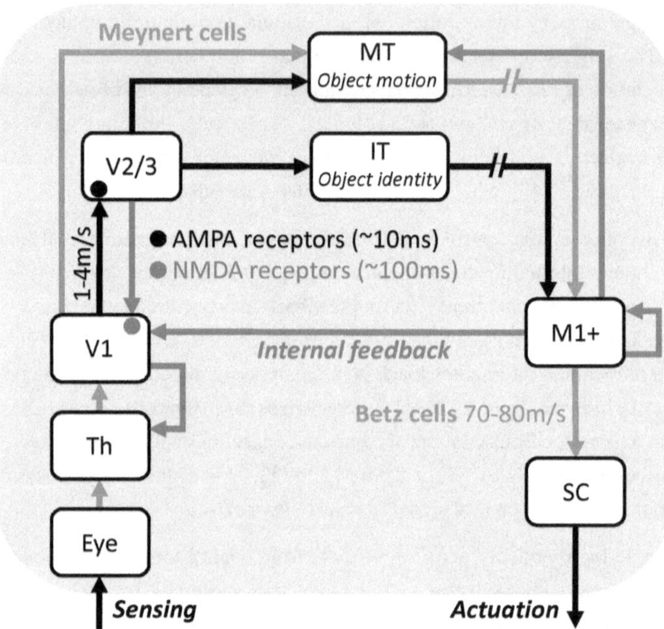

Figure 6.2: A partial, simplified schematic of sensorimotor control. We focus on key cortical and subcortical areas and communications between them. Black and green arrows indicate communications that traverse from sensing toward actuation; green arrows are particularly fast pathways, which enable the tracking of moving objects in our model. Pink arrows indicate internal feedback signals, which traverse from actuation toward sensing. Broken lines are not necessarily direct neuronal projections. SC = spinal cord, Th = thalamus, V1 = primary visual cortex, M1+ = primary motor cortex and additional motor areas, V2/3 = secondary and tertiary visual cortex, IT = inferotemporal cortex, MT = mediotemporal cortex (V5). Only a subset of the internal feedback pathways are shown (e.g. not included are internal feedback signals from M1+ to V2 and signals from M1+ to IT).

itively, signals related to movements of the whole body are found in areas typically associated with particular parts of the body, such as the hand area, as well as sensory areas such as primary visual cortex [56]–[59]. Indeed, recent analysis of the correlation structure between neurons during a visual discrimination task revealed a task-related global mode in the correlations between cortical neurons associated with the task response rather than the sensory stimulus, strongly supporting the idea that top-down feedback is an important element of sensory processing [60]. These motor-related signals in sensory pathways, which span subsystems and tasks, are generated by internal feedback and are the focus of this study.

Internal feedback has been studied in the context of predictive coding [61], [62] and have been invoked in other modeling studies [63]–[66]. However, these models focus on sensory or motor systems separately and do not account for key constraints on neuronal communication in both space and time to achieve sensorimotor tasks. Achieving fast and accurate computation and communication over across brain areas is difficult, or even impossible, because communication may be slow, limited in bandwidth and constrained to spatially localized populations.

Here, we build on the foundations of recent work in distributed control theory [3], [67]–[70] and show that internal feedback is a solution to achieving rapid and accurate control given the spatial and temporal constraints on brain components and communication systems. We analyze an idealized class of control models and prove mathematically that internal feedback is both plausible and *necessary* for achieving optimal performance in these idealized models. Internal feedback serves at least three functions in our model: *prediction and estimation, localization of function and behavior*, and *focused attention*, all of which are crucial for effective sensorimotor control and survival. This theory explains why there are differences in population responses between M1 and V1, why different projections predominantly activate AMPA or NMDA glutamate receptors, the functions of giant pyramidal cells in visuomotor control, and both the uses and limitations of localization of function in cortex. There is a general principle behind all of these physiological properties.

6.2 Task Model and Performance

We analyze expected values and theoretical bounds on task performance for highly simplified control loop models motivated by a well-studied and ethologically relevant *tracking* task: reaching for a moving object. The goal of the task is continuous pursuit, rather than one-time contact between limb and object.

The complete tracking task requires identification of the object in a cluttered visual scene, prediction of the object's movement, and generation and execution of bimanual limb movement. We make many simplifying assumptions that allow us to study internal feedback in an accessible way using familiar linear dynamical systems. Models of greater detail and complexity are discussed briefly and are expected to be based on the same broad principles.

Consider the task of tracking a moving object with the endpoint of a limb on a plane. The variable to be controlled is the *tracking error*: the distance between the hand and the object. We start by assuming that the system controlling the limb can perfectly sense the position of limb and object at every instant, which will be relaxed in later models. The *cost* is defined as the squared Euclidean norm of the tracking error over time, normalized by the total amount of time; with a smaller cost indicating better tracking.[2].

Let x, u, and w represent the tracking error, the control action on the limb, and the action of the object, respectively. We will refer to x as the *state* of the system. Let A be a matrix that represents the intrinsic dynamics of x, including features such as the movement of the object or mechanical coupling between two dimensions of limb movement. Let B be a matrix that represents the mapping of control action to tracking error. The time-evolving dynamics of the tracking error follows from the linear equation of motion (2.1).

Let α denote the magnitude of the maximum eigenvalue of A, as a proxy for task difficulty. Note that $\alpha < 1$ corresponds to a task in which tracking error x will decrease with no limb action, an easy task. In general, the difficulty of a task will depend on properties of A such as its eigenvalues and the strength of coupling between states, as discussed in subsequent sections. For example, if the spectral radius of A is less than 1, this corresponds to a task in which tracking error x decreases with no limb action, an easy task.

The actions u provide feedback control on the tracking error, computed by an arbitrary function $\mathcal{K}$ that has access to all past and present tracking errors $x(1 : t)$. This control law is mathematically written as

$$u(t) = \mathcal{K}(x(1 : t)) \tag{6.1}$$

---

[2]For control theorists: this corresponds to a unit state penalty and infinitesimal control penalty.

The optimal solution to this problem is the linear quadratic regulator (LQR) and the optimal controller is $\mathcal{K}(x(1:t)) = -Ax(t)$ [43]. This controller fits into the single-loop model of sensorimotor control, as there is no internal feedback, and the addition of internal feedback does not provide any additional performance advantage. We also note that even though we allow access to all past values of x, the optimal controller only needs to access x at time t.

Controllers without internal feedback are optimal for a large but special class of problems, including classical state feedback and full control problems from control theory. Though mathematically elegant, these controllers make assumptions that are impractical when applied to biological systems. In subsequent sections, we relax some of the assumptions implicit in this single-loop model and show that small deviations from assumptions relevant to biological systems introduces the need for internal feedback.

Any of the controllers in subsequent sections can be implemented in a variety of ways, although whether or not a particular controller needs internal feedback is generic across all possible implementations. We choose particular non-unique controller implementations with internal feedback for which the optimal solution is relatively transparent and easy to interpret.

6.3 Sensing and Actuation Delay

The linear quadratic regulator and linear quadratic Gaussian controller are two of the most widely-used optimal controllers throughout academia and industry. However, these formulations lack the ability to incorporate delayed sensing and delayed actuation, which are prevalent in organisms. For instance, the neuronal conduction time from the eye to motor cortex, has conduction delays on the order of tens or hundreds of milliseconds. In this section, we provide new techniques that allow us to incorporate delays into these standard formulations, then investigate the internal feedback within the resulting controllers.

Full control and state feedback

The state feedback problem is one in which we assume direct access to the state: given dynamics (2.1), we have the feedback law

$$u(t) = Kx(t) \tag{6.2}$$

and we are tasked with finding the optimal controller K, which is a constant matrix. State feedback can be thought of as a controller with perfect sensing and arbitrary actuation. Its dual, the full control problem, can be thought of as a controller with perfect actuation and arbitrary sensing[3]. In full control, we assume that we work with dynamics (2.1) with actuation matrix $B = I$. The controller is

$$u(t) = Ly(t) \tag{6.3}$$

where $y(t)$ represents sensor data and is related to the state through

$$y(t) = Cx(t) \tag{6.4}$$

We will first formulate the delayed sensing problem in full control. We have a sensor with a single timestep of delay, written as

$$y(t + 1) = Cx(t) \tag{6.5}$$

and we want to find the best controller (6.3) for it. To model sensing delay, we add delay states to the model. Assume that we sense each state, but with a single timestep of delay. We introduce a virtual internal state x_s, which contains delayed information about state x. This formulation allows us to pose the delayed-sensor tracking problem as a standard control problem

$$\tilde{A} = \begin{bmatrix} A & 0 \\ C & 0 \end{bmatrix}, \quad \tilde{C} = \begin{bmatrix} 0 & I \end{bmatrix}$$
$$\begin{bmatrix} x(t+1) \\ x_s(t+1) \end{bmatrix} = \tilde{A} \begin{bmatrix} x(t) \\ x_s(t) \end{bmatrix} + u(t) + \begin{bmatrix} w(t) \\ 0 \end{bmatrix} \tag{6.6}$$

This problem can be optimally solved by LQR. The optimal controller L can be found by solving a discrete-time algebraic Riccati equation on $\tilde{A}$ and $\tilde{C}$.

Control action u and controller L can be further partitioned as

$$\begin{bmatrix} u_1(t) \\ u_2(t) \end{bmatrix} = \begin{bmatrix} L_1 \\ L_2 \end{bmatrix} y(t) \tag{6.7}$$

where L_1 and L_2 are block matrices. Here, the controller does not directly "perceive" the tracking error x and only has access to the virtual internal state x_s. However, the controller can freely take actions that affect both the tracking error and the virtual state. The action on the virtual state, u_2, as shown in Figure 6.4, is an example of counterdirectional internal feedback with gain L_2. We remark that even

---

[3]This formulation is less often used, as it is much more common to have perfect sensing than perfect actuation.

though u_1 and u_2 are both control signals in the standard sense, they have vastly different interpretation in this model: u_1 represents physical actuation, which for an organism requires high-cost muscle cells, etc, while u_2 represents low-cost internal communication, i.e. neurons.

For the delayed sensing problem, the optimal controller has a simple analytical form: $L_1 = -A^2$ and $L_2 = -A$ is the internal feedback. If no internal feedback is allowed (i.e. we enforce $L_2 = 0$), then the optimal controller is $L_1 = -A^2/4$. We compare the performance of these two controllers in Figure 6.3, and see that the controller with internal feedback far outperforms the controller without internal feedback. We also note that as the task becomes more difficult (spectral radius of $A > 2$), the controller without internal feedback is unable to stabilize the closed-loop system and tracking breaks down.

We can also pose the dual problem of delayed actuation as a standard control problem, in the form of

$$\tilde{A} = \begin{bmatrix} A & B \\ 0 & 0 \end{bmatrix}, \quad \tilde{B} = \begin{bmatrix} 0 \\ I \end{bmatrix}$$

$$\begin{bmatrix} x(t+1) \\ x_a(t+1) \end{bmatrix} = \tilde{A} \begin{bmatrix} x(t) \\ x_a(t) \end{bmatrix} + \tilde{B}u(t) + \begin{bmatrix} w(t) \\ 0 \end{bmatrix} \tag{6.8}$$

$$u(t) = \begin{bmatrix} K_1 & K_2 \end{bmatrix} \begin{bmatrix} x(t) \\ x_a(t) \end{bmatrix}$$

Here, control signal u is delayed before being applied to state x. Virtual state x_a represents the delayed actuation signal. The optimal controller K can be found by solving a discrete-time algebraic Riccati equation on $\tilde{A}$ and $\tilde{B}$. Here, K_2 represents internal feedback, as shown in shown in Figure 6.4.

The remainder of this subsection is intended primarily for readers with some background in control theory. We present the full mathematical details for a system with more than one step of delay – we will do this for the case of full control, and the state feedback case follows naturally by duality.

Consider a system of size n, with sensory input of size p and sensory delay of d timesteps. As before, we will introduce virtual states x_s, this time with dimension pd. The resulting state vector is $\tilde{x} = \begin{bmatrix} x^\mathsf{T} & x_s^\mathsf{T} \end{bmatrix}^\mathsf{T}$, where x is the original tracking error. We can think of x_s as information sensed by the eye that is being passed along a delayed visual pathway to LGN, then V1, and so on. The motor areas of the brain

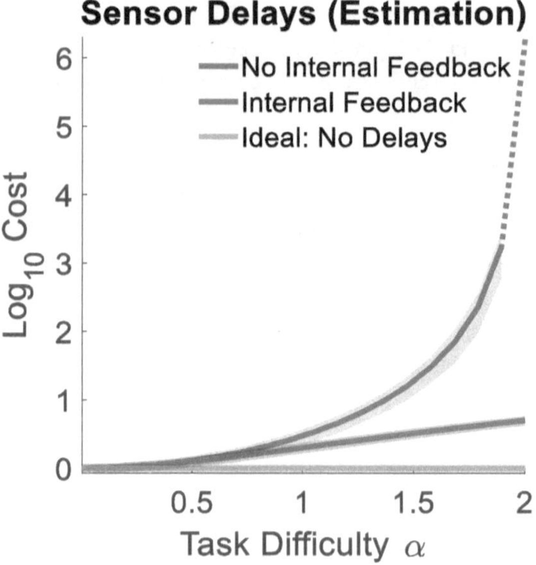

Figure 6.3: Internal feedback improves performance when there are internal delays in sensing. The scalar problem of tracking a moving target over a line was simulated, varying the task difficulty . The 'Ideal' controller contains no sensor delays. The 'Internal Feedback' controller contains sensor delays, and uses internal feedback to compensate for the delays. The 'No Internal Feedback' controller contains sensor delays, but uses no internal feedback. As α approaches 2, the task becomes infeasible without internal feedback (broken line). Shaded areas indicate standard deviations.

can only access the most delayed information, i.e. the values of x_s corresponding to a delay of d; these will be picked out by the sensing matrix C. Thus, the overall setup is

$$\tilde{A} = \begin{bmatrix} A & & 0_{n\times pd} \\ C & & 0_{p\times pd} \\ 0_{p(d-1)\times n} & I_{p(d-1)} & 0_{p\times p} \end{bmatrix}, \quad \tilde{C} = \begin{bmatrix} 0_{p\times n+p(d-1)} & I_p \end{bmatrix},$$

$$(6.9)$$

$$\begin{bmatrix} x(t+1) \\ x_s(t+1) \end{bmatrix} = \tilde{A} \begin{bmatrix} x(t) \\ x_s(t) \end{bmatrix} + u(t) + \begin{bmatrix} w(t) \\ 0 \end{bmatrix}$$

As before, the optimal control law is (6.3), and the optimal controller L can be found by solving a discrete-time algebraic Riccati equation on $\tilde{A}$ and $\tilde{C}$. We can

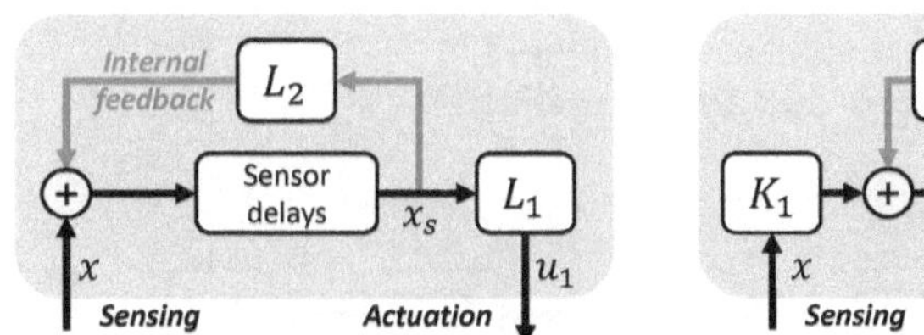

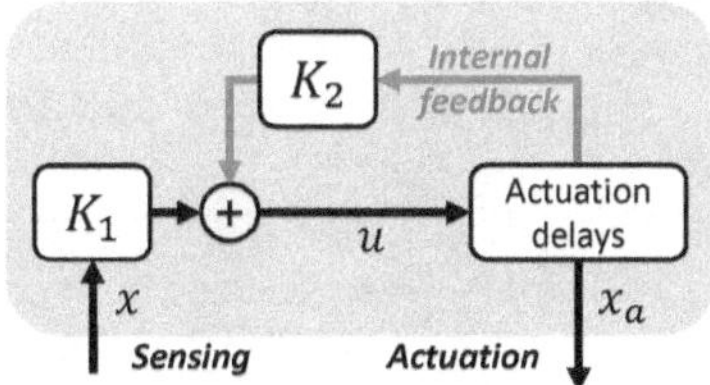

Figure 6.4: **(Left)** Optimal control model for system with sensor delays. Tracking error x is sensed, then communicated by the sensor with some delay to the L_1 block, which computes the appropriate actuation. Counterdirectional internal feedback (pink) conveys information from actuation back toward sensing. Internal computation L_2 adjusts the sensor signal to compensate for actions taken by the system; this results in improved performance. **(Right)** Optimal control model for system with actuation delays. Actuation u is computed, then executed with some delay. Counterdirectional internal feedback (pink) conveys information back toward sensing, to compensate for the actuation delay.

once more partition the controller L as is done in (6.7); now, u_2, which represents internal feedback, has dimension pd. As we can see, the dimension of this internal feedback grows with delay d. We show a depiction of a system with three timesteps of sensory delay in Figure 6.5, and a system with three timesteps of actuation delay in Figure 6.6. These diagrams can be considered to be more zoomed-in versions of 6.4. Bracketed values indicate partitions of blocks, i.e.

$$L_2 = \begin{bmatrix} L_2(1) \\ L_2(2) \\ L_2(3) \end{bmatrix} \tag{6.10}$$

where each submatrix $L_2(i)$ has n rows. Similarly,

$$K_2 = \begin{bmatrix} K_2(1) & K_2(2) & K_2(3) \end{bmatrix} \tag{6.11}$$

where each submatrix $K_2(i)$ has n columns.

In both delayed sensing and delayed actuation, internal feedback is required for optimal performance. Internal feedback adjusts delayed signals to compensate for actions taken and information received during the delay; in other words, internal feedback implicitly compensates for the delays.

The linear-quadratic problem that we consider here could be straightforwardly tested in a laboratory setting. The simplest real-world task, which is exactly captured by

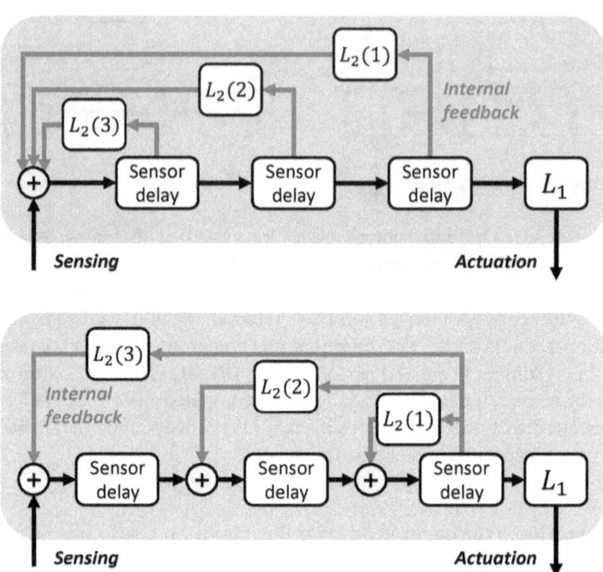

Figure 6.5: **(Top)** Optimal control model for system with 3 steps of sensor delay. Counterdirectional internal feedback (pink) conveys information from actuation back toward sensing. Internal computations L_2 adjusts the sensor signal to compensate for actions taken by the system; this results in improved performance. **(Bottom)** Alternative equivalent implementation of optimal control model for system with 3 steps of sensor delay.

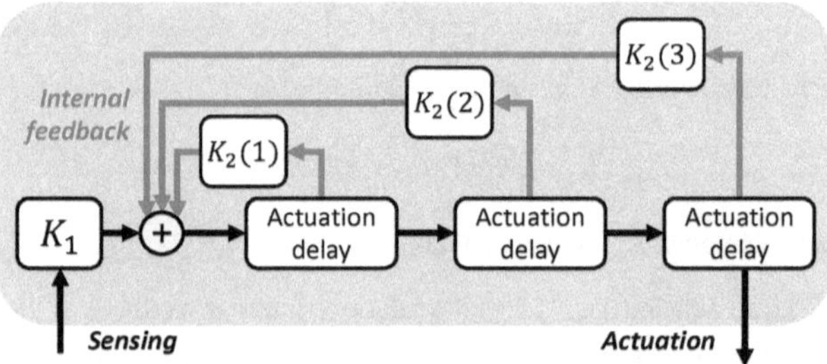

Figure 6.6: Optimal control model for system with 3 steps of actuation delays. Actuation u is computed, then executed with some delay. Counterdirectional internal feedback (pink) conveys information back toward sensing, to compensate for the actuation delay.

the model, is a stable limb resting on a surface tracking an object over a line or plane. This can be modeled by $A = I$ with single time-step delays corresponding to internal loop delays on the order of 100 ms. Longer delays would move the feasibility boundary towards a spectral radius of 1. More complex models of more difficult tasks corresponding to more realistic scenarios, such as movement against gravity and adversarial disturbances, could make a task impossible to control without internal feedback. We next describe controllers with internal feedback that include sensor delays, actuator delays and imperfect sensing that will motivate a general model for sensorimotor control in brains.

Kalman filter with and without delays

We now consider the case in which sensing is instantaneous, but noisy. Consider dynamics (2.1) with sensor (6.4) — this is the output feedback control problem. The optimal controller makes use of controller gain K and estimator gain L, and is formulated as

$$\hat{x}(t + 1) = A\hat{x}(t) + Bu(t) + L(y(t) - C\hat{x}(t))$$
$$u(t) = K\hat{x}(t)$$

(6.12)

where $\hat{x}$ is an internal estimate of tracking error x. This optimal controller uses the Kalman filter, which inherently contains three counterdirectional internal feedback pathways irrespective of delays being present. These pathways are represented by the blue arrows through in Figure 6.7, and play a central role in state estimation. The pathway through A estimates state evolution in the absence of noise and actuation; the pathway through B accounts for controller action, and the pathway through C predicts incoming sensory signals based on the internal estimated state.

The implementation shown in Figure 6.7 is not unique. We now briefly discuss a few equivalent implementations that use less internal feedback and explain why they are less advantageous in the sensorimotor context than the implementation in Figure 6.7.

In one alternative implementation, we can remove the internal feedback through B, and replace A with $A + BK$. However, this requires duplication of K; in the sensorimotor context, which requires duplicating of motor structures within visual structures. In another alternative implementation, we can remove the internal feedback through C, and replace A with $A - LC$. This requires a duplication of L. Additionally, filtering out predictable sensory input via C earlier (as is done in Figure 6.7) can be preferable to filtering it out later (as in our alternative implementation). This is because the filtered information is typically much smaller in bandwidth, and

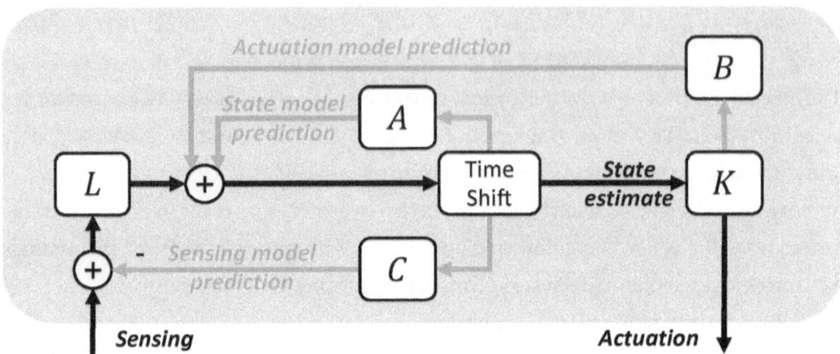

Figure 6.7: Internal feedback in a controller with instantaneous but imperfect sensing and actuation. A, B, and C represent the state, actuation, and sensing matrices of the physical plant; K represents the optimal controller, and L represents the optimal observer. The Time Shift block shifts $\hat{x}(t+1)$ to $\hat{x}(t)$ in Eq. 5. The internal feedback pathways (blue) are inherent to the Kalman Filter; these use state, actuation, and sensing models to create an internal estimate of the tracking error, or state. All internal feedback depicted in this diagram is counterdirectional.

requires less resources to communicate: the earlier we perform this filtering, the less resources we require to pass this information forward. If communications are subject to a speed-accuracy trade-off (described below), then earlier filtering allows us to pass sensory information forward with less delay.

We now synthesize a model that combines features from previous sections: sensor delays, actuator delays, and imperfect sensing. The model can be constructed using virtual states as follows, for the case of one timestep of delay

$$
\begin{bmatrix} x(t+1) \\ x_a(t+1) \\ x_s(t+1) \end{bmatrix} = \begin{bmatrix} A & B & 0 \\ 0 & 0 & 0 \\ C & 0 & 0 \end{bmatrix} \begin{bmatrix} x(t) \\ x_a(t) \\ x_s(t) \end{bmatrix}
$$
$$
+ \begin{bmatrix} 0 & 0 \\ I & 0 \\ 0 & I \end{bmatrix} \begin{bmatrix} u(t) \\ u_s(t) \end{bmatrix} + \begin{bmatrix} w(t) \\ 0 \\ 0 \end{bmatrix} \tag{6.13}
$$
$$
y(t) = x_s(t)
$$

where x_a and x_s are virtual internal states corresponding to delayed actuator commands and delayed sensor signals, respectively, and u_s represents compensation on virtual internal states. We can use standard control theory to obtain the optimal

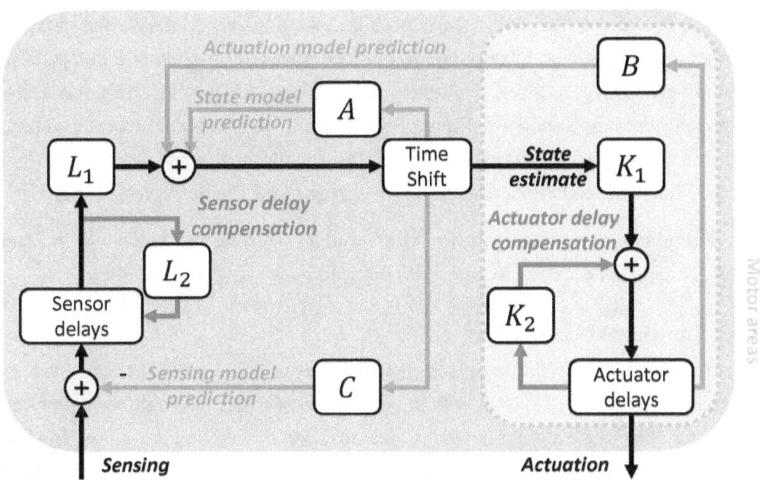

Figure 6.8: Internal feedback in a controller with sensor and actuator delays. A, B, and C represent the state, actuation, and sensing matrices of the physical plant; K_1, K_2, L_1, L_2 are submatrices of the optimal controller and observer gains. The internal feedback pathways (pink) through L_2 and K_2 compensate for sensor and actuator delays, respectively. Other internal feedback pathways (blue) are inherent to the Kalman Filter. All internal feedback depicted in this diagram is counterdirectional. The yellow box contains parts of the controller that roughly correspond to motor areas in cortex.

controller gain K and optimal estimator gain L. Due to the block-matrix structure of the system matrices, the optimal gains have the following structure

$$K = \begin{bmatrix} K_1 & K_2 & 0 \end{bmatrix}, L = \begin{bmatrix} L_1 \\ 0 \\ L_2 \end{bmatrix} \tag{6.14}$$

The controller can be implemented as follows

$$\delta(t+1) = Cx(t) - C\hat{x}(t) - L_2\delta(t)$$
$$\hat{x}(t+1) = A\hat{x}(t) + Bx_a(t) + L_1\delta(t) \tag{6.15}$$
$$u(t) = K_1\hat{x}(t) + K_2x_a(t)$$

Here, δ is the delayed difference between the estimated sensor input and true sensor input, discounted by the observer term $L_2\delta(t)$. The resulting controller, shown in

Figure 6.8, contains two internal feedback pathways related to delay; one pathway compensates for sensor delays, and the other compensates for actuator delays. The remaining internal feedback is inherent to the Kalman Filter, as described in the previous section and shown in Figure 6.7. Overall, the inclusion of sensor delays, actuator delays, and imperfect sensing result in an optimal controller with several internal feedback pathways, each of which serves a specific, interpretable purpose.

We additionally remark that in the adversarial $\mathcal{H}_\infty$ setting, an additional internal feedback path is necessary to anticipate the worst-case adversarial disturbance.

6.4 Functional Localization

Almost all muscles in the body are engaged in even the simplest actions, such as reaching. Controlling a system with many degrees of freedom is a difficult problem for motor control even without delays. Localization of function is well established in motor cortex, with different different body parts controlled by different cortical regions; however, communication and computation between localized cortical areas typically has spatial and temporal constraints, compared with signals within areas.

Consider two motor areas and partition tracking errors into two sets x_1 and x_2, representing two distinct but possibly coupled subsystems (e.g. two distinct limbs that are mechanically coupled) using the problem formulation described by (2.1). The overall tracking error is

$$x = \begin{bmatrix} x_1 \\ x_2 \end{bmatrix} \tag{6.16}$$

Correspondingly, we partition actuators into two sets u_1 and u_2 that act on their respective subsystems via local controllers

$$u = \begin{bmatrix} u_1 \\ u_2 \end{bmatrix} \tag{6.17}$$

Each local controller senses and controls one subsystem; i.e. local controller 1 senses x_1 and computes u_1, and local controller 2 senses x_2 and computes u_2. Local controllers may communicate with one another; however, due to localization constraints, the cross-communication is delayed. Thus, local controller 1 cannot directly access x_2 without some delay, and similarly for local controller 2.

We observe that without the constraint of localized communication, the optimal controller for our system is $u = -Ax$. If A is block-diagonal (i.e. x_1 and x_2 are uncoupled), then this controller obeys localized communication constraints —

in fact, no cross-communication (internal feedback) is required between the two local controllers. However, if the two subsystems are coupled with time delays, then this controller requires rapid, global communication, which violate localized communication constraints.

To enforce localized communication, we reformulate the problem by introducing virtual states x_1' and x_2', which represent delayed cross-communication between the two local controllers. x_1' is information sent from local controller 1 to local controller 2, with delay. x_2' is information sent from local controller 2 to local controller 1, with delay. We also define u_1' and u_2', which model interconnections between virtual states and real tracking errors. For simplicity, we assume unit delay. The reformulated problem then becomes

$$\tilde{x} = \begin{bmatrix} x_1 \\ x_2' \\ x_1' \\ x_2 \end{bmatrix}, \tilde{u} = \begin{bmatrix} u_1 \\ u_2' \\ u_1' \\ u_2 \end{bmatrix} \tilde{w} = \begin{bmatrix} w_1 \\ 0 \\ 0 \\ w_2 \end{bmatrix},$$

$$\tilde{A} = \begin{bmatrix} A_{11} & 0 & 0 & A_{12} \\ 0 & 0 & 0 & 0 \\ 0 & 0 & 0 & 0 \\ A_{21} & 0 & 0 & A_{22} \end{bmatrix}, K = \begin{bmatrix} * & \triangle & * & 0 \\ \triangle & * & \triangle & * \\ * & \triangle & * & \triangle \\ 0 & * & \triangle & * \end{bmatrix} \tag{6.18}$$

$$\tilde{x}(t+1) = \tilde{A}\tilde{x}(t) + \tilde{u}(t) + \tilde{w}(t)$$

$$\tilde{u} = K\tilde{x}$$

The zeros in the top right and bottom left corners of the K matrix preserve localized communication; they enforce that the two local controllers cannot communicate instantaneously to one another. Asterisks and triangles indicate free values: triangles represent sites of potential cross-communication, or lateral internal feedback. When these free values are optimized to achieve optimal performance with localized communication, the resulting K matrix is

$$K = \begin{bmatrix} -A_{11} & 0 & -A_{12} & 0 \\ 0 & 0 & -A_{12} & A_{12} \\ A_{21} & -A_{21} & 0 & 0 \\ 0 & -A_{21} & 0 & -A_{22} \end{bmatrix} \tag{6.19}$$

The resulting local controllers are shown in Figure 6.9. Note that the $-A_{12}$ term in the second row and the $-A_{21}$ term in the fourth row of K correspond to lateral

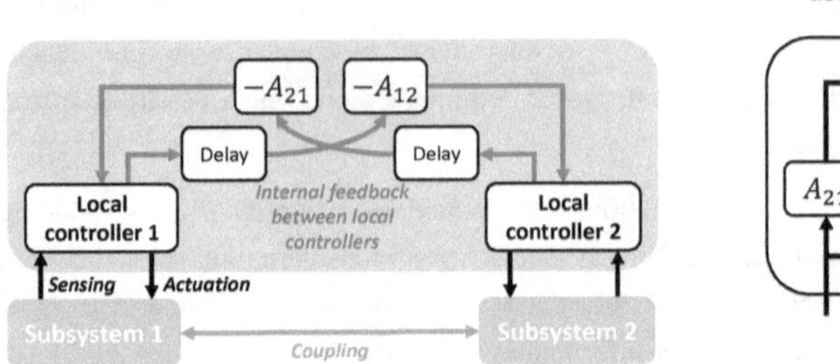

Figure 6.9: Optimal localized control of two coupled subsystems. **(Top)** Overall schematic. Each subsystem has its own corresponding local controller, which senses and actuates only its assigned subsystem. Local controllers communicate to each other via lateral internal feedback (pink), with some delay. **(Bottom)** Circuitry of local controller 1. Local controller 2 has identical circuitry, with different matrices; A_{12} instead of A_{21}, A_{22} instead of A_{11}, etc.

internal feedback. Here, these internal feedback signals carry predicted values of the unsensed tracking errors for each controller, after taking control action into account; for instance, internal feedback from local controller 2 to local controller 1 conveys the predicted value of x_2, after taking control action from controller 2 into account. We can develop intuition for this implementation by following an impulse w through time

$$\tilde{x}(1) = \begin{bmatrix} w_1 \\ 0 \\ 0 \\ w_2 \end{bmatrix} \rightarrow \tilde{x}(2) = \begin{bmatrix} A_{12}w_2 \\ A_{12}w_2 \\ A_{21}w_1 \\ A_{21}w_1 \end{bmatrix} \rightarrow \tilde{x}(3) = 0 \tag{6.20}$$

The performance of this controller can be compared to the controller without internal feedback in Figure 6.10. The best possible linear controller for the controller without internal feedback results in severe performance degradation. As task difficulty increases, this controller is unable to stabilize the closed-loop system and tracking becomes infeasible. With internal feedback, task performance stays near the centralized optimal (i.e. the case where local controllers can communicate freely without delay).

This analysis shows that when motor function is localized to specialized parts of motor cortex that control particular parts of the body, cross-communication via internal feedback between local controllers is essential. Local circuits in the two

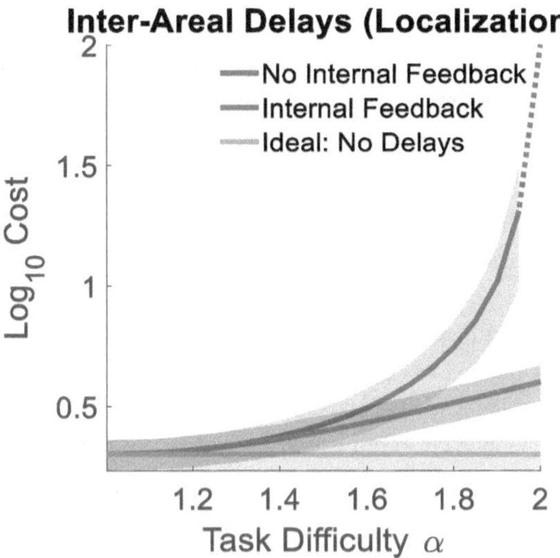

Figure 6.10: Localization of function within motor-related cortex: although different parts of the cortex control different parts of the body, these parts of the body are inherently mechanically coupled. As a result, internal feedback is useful and in some cases necessary to maintain localization of function. In simulations, we consider the problem of tracking a moving target over a two-dimensional space, varying the task difficulty. The "Ideal" controller is centralized (i.e. no delays between local controllers) and obtains the best performance. The localized controller with internal feedback achieves similar performance. The localized controller without internal feedback suffers from substantially worse performance (higher cost). As task difficulty increases, the task becomes infeasible without internal feedback (broken line). Shaded areas indicate standard deviations.

hemispheres must also be coordinated — indeed, they are connected by a massive corpus collosum that crosses the midline.

The crosstalk between local controllers is supported by the presence of global signals from movements of the whole body to the local controller in motor cortex specialized to particular parts of the body. In reality, all body movements are mechanically coupled, something which the motor system can conceal through effective localization and coordination using internal feedback.

6.5 Communication Constraints and Distributed Sensorimotor Circuits

The above examples use tools from standard control theory to model delays and local communication constraints in controllers. We now use tools from SLS to model both delays and local communication constraints in a slightly more complex system. This section is geared toward readers with some background in control theory.

We consider a chain of coupled subsystems (shown in the top left of Figure 6.12) which is controlled in a distributed fashion, and whose distributed controller is subject to delays and local communication constraints. This distributed controller, which can be thought of as a model of distributed sensorimotor circuitry, contains large amounts of internal feedback — both counterdirectional and lateral. This is the first model that replicates the massive amounts of internal feedback observed in the cortex from purely *a priori* principles. In our case, distributed implementation, communication constraints, and sensing-to-actuation ratio are the principles which drive the presence of internal feedback.

Though we provided examples of how classical control can be modified to accommodate delays and local communication, we will be utilizing the SLS parametrization in this section. The primary reason is that delays and local communication, while simple to formulate on small systems, become unwieldy (and, in the case of local communication, downright impossible) to solve on larger systems using the tools of classical control. All previous examples can be equivalently solved using SLS.

We use a symmetric 8-subsystem ring, shown in the top left of Figure 6.12. The state matrix $A \in \mathbb{R}^{8 \times 8}$ and actuation matrix $B \in \mathbb{R}^{8 \times 4}$ are

$$A = \frac{\alpha}{3} * \begin{bmatrix} 1 & 1 & 0 & \dots & 1 \\ 1 & 1 & 1 & 0 & \dots \\ 0 & \ddots & \ddots & \ddots & \vdots \\ \vdots & \ddots & \ddots & \ddots & 1 \\ 1 & 0 & \dots & 1 & 1 \end{bmatrix}, B = \begin{bmatrix} 1 & 0 & 0 & 0 \\ 0 & 0 & 0 & 0 \\ 0 & 1 & 0 & 0 \\ \dots & \dots & & \\ 0 & 0 & 0 & 0 \end{bmatrix} \qquad (6.21)$$

where the spectral radius of A is equal to $\alpha = 1.8$. This system is unstable; in context of the tracking task, it is relatively difficult. 50% of subsystems are actuated; subsystems 1, 3, 5, and 7 receive actuation, while subsystems 2, 4, 6, and 8 do not.

The standard implementation of the SLS controller is shown in Figure 2.2. We now look at the distributed implementation of this controller. For pedagogical reasons, we will first focus on the implementation of the $z\Phi_u$ block alone, then discuss the

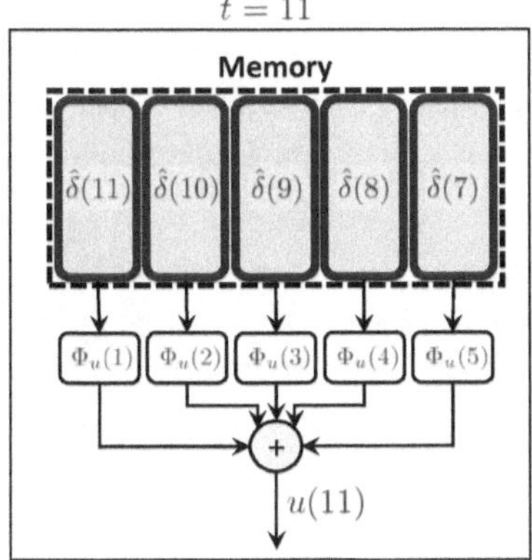

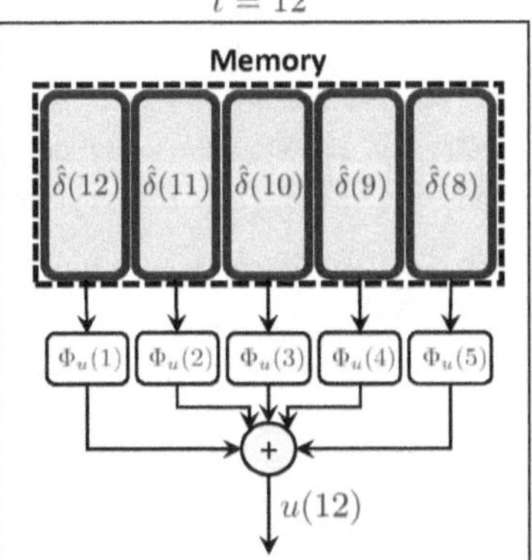

Figure 6.11: Standard implementation of a system with FIR transfer function Φ_u, input $\hat{\delta}$, and output **u**. This example uses a strictly causal Φ_u with horizon length $T = 5$. Values in memory are multiplied by the appropriate spectral element $\Phi_u(k)$ then summed to obtain the output. **(Left)** Contents of memory at time $t = 11$. **(Right)** Contents of memory at time $t = 12$. Notice that the entries in memory have shifted to the right compared to the previous timestep. The oldest value, $\hat{\delta}(7)$ has been discarded, and a new value, $\hat{\delta}(12)$, has entered the memory.

implementation of the full controller. This block takes $\hat{\delta}$ as input and gives **u** as output. In time domain, this block performs the following convolution:

$$u(t) = \sum_{k=0}^{T} \Phi_u(k)\hat{\delta}(t - k + 1) \tag{6.22}$$

To compute $u(t)$, we require a memory of the most recent T values of $\hat{\delta}$. Every time step, we discard the oldest value of $\hat{\delta}$ from memory and add the newest value of $\hat{\delta}$. An example of this is shown in Figure 6.11. This is a standard implementation of a system with FIR transfer matrix Φ_u; we will build upon it to create locally implemented controllers with local memory.

To incorporate both delay and local communication constraints, we enforce sparsity on transfer matrix Φ_u. This corresponds to sparse spectral elements $\Phi_u(k)$. These sparsity constraints enable us to implement this controller in a distributed manner: instead of a centralized memory and convolution computation, each subsystem will have its own local memory and local convolution computation. Each subsystem i locally computes its input $\hat{\delta}_i$, and relies on local inter-subsystem communication to access $\hat{\delta}_j$ for $j \neq i$. We show an example in Figure 6.12, where each subsystem communicates only with its immediate neighbors. In our example, we consider inter-subsystem delays, but not self delays (i.e. sensor i communicates to controller i with some nonzero delay). Self delays can be easily enforced via zero-constraints on the diagonals of $\Phi_u(k)$ for appropriate k.

Since each subsystem has its own local memory, a lot of redundant memory is created. In our example, the memory at subsystem 4 stores past values of $\hat{\delta}_3$; past values of $\hat{\delta}_3$ are also stored at subsystems 2 and 3. Here, each entry in memory is stored 3 times: by subsystem i and its neighbors $i + 1$ and $i - 1$. For a general system, each entry will be stored by subsystem i and all neighbors with whom it communicates.

Figure 6.12 shows an example of local memory at a single subsystem. A more general characterization is shown in Figure 6.13. In our example, the size of the local memory is uniform across subsystems; however, in general, the size of the local memory may vary from subsystem to subsystem. All of this is supported by the SLS formulation; one must only specify the appropriate sparsity constraint on Φ_u and Φ_x.

We have described the implementation of the $z\Phi_u$ block from Figure 2.2. To implement the full SLS controller described in this figure, we must also implement the $z\Phi_x - I$ block. This block also takes $\hat{\delta}$ as input, so it shares the memory structure we have already described. We now present the full local controller at a single subsystem in Figure 6.14. Here, the counterdirectional internal feedback computes $\hat{x}$, the predicted future state. It does so using the closed-loop map Φ_x, which contains information on both the system itself *and* the system's controller (i.e. motor model). This echoes observations from previous sections and existing literature: counterdirectional feedback serves a predictive purpose. We also note that in state feedback, the "estimations" are exact; $\hat{\delta}(t) = w(t-1)$, and $\hat{x}(t+1) = x(t+1)$ if $w(t) = 0$.

Lateral feedback is also abundant in the local controller. As in the previous section, it serves the purpose of cross-communication between sub-controllers of subsystems that are dynamically coupled.

To get a better picture of the different sources of internal feedback, we show three local controllers in Figure 6.15. Here, subsystem 4 is unactuated, as per the original problem statement above. However, despite having no actuation, subsystem 4 still

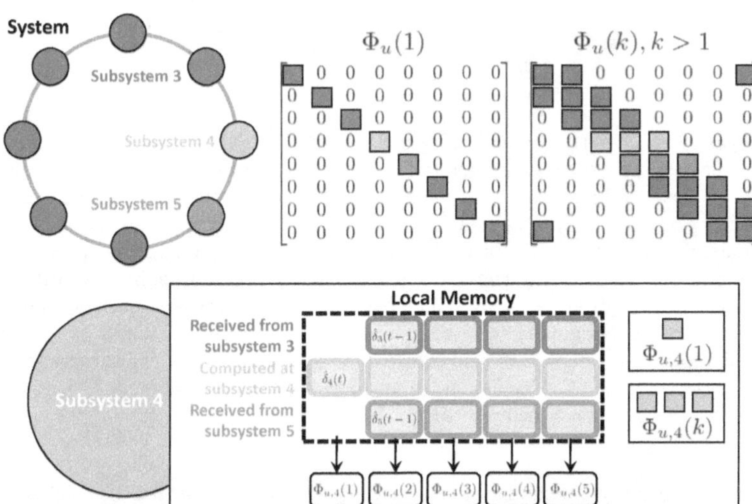

Figure 6.12: For ease of visualization, we temporarily assume all subsystems are actuated. **(Top left)** Ring with 8 subsystems. **(Top right)** Spectral elements of Φ_u. Colored squares represent nonzero values; other values are constrained to be zero. Nonzero values away from the diagonal represent communication between subsystems. Sparsity constraints arise from delayed communication (for $\Phi_u(1)$) and local communication (for $\Phi_u(k), k > 1$). Sparsity on Φ_u additionally translates to local disturbance rejection. **(Bottom)** Local controller and memory at subsystem 4. Each subsystem uses its own row of $\Phi_u(k)$ to implement its local controller. Rectangles in local memory represent scalar values of $\hat{\delta}_i(t)$; colors indicate the source of the value, e.g. pink rectangles are $\hat{\delta}$ values from subsystem 3. Recent entries are toward the left, and oldest entries are toward the right. Local actuation (not shown) is produced by multiplying $Phi_{u,4}(k)$ by columns in memory and summing over the products.

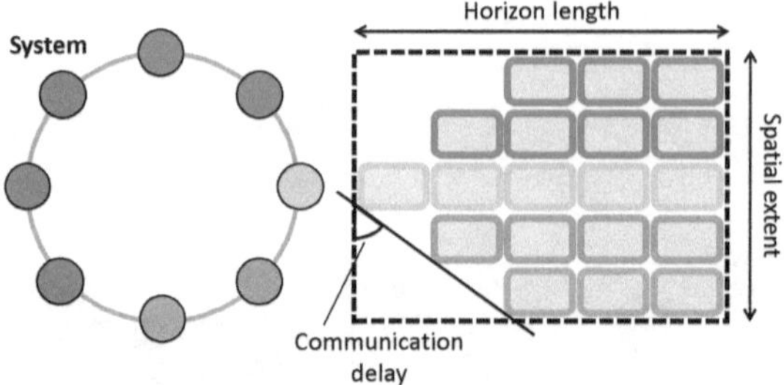

Figure 6.13: **(Left)** Ring with 8 subsystems. **(Right)** Size and shape of example local memory at subsystem 4. Horizon length T indicates how far to remember into the past. Spatial extent indicates how many neighbors each subsystem communicates with. For the ring system, communication delay is indicated by the angle of the triangular "front" of the memory; larger angle corresponds to larger delay. In this example, $T = 5$, and subsystems communicate to their 4 nearest neighbors with delay proportional to distance. Subsystem 4 has up-to-date information on subsystem 4 (yellow), slightly outdated (delayed one time step) information from subsystems 3 and 5 (pink and green), and more outdated (delayed 2 time steps) information from subsystems 2 and 6, which are farther away. Note that for an arbitrary system, the pattern of local memory entries will not be triangular.

requires memory and circuitry to calculate $\hat{\delta}_4$ and communicate it to its neighbors. A fundamental asymmetry between feedforward and internal feedback is revealed here; in our model, feedforward is necessitated by actuation, while internal feedback is necessitated by sensing. In both biological and cyberphysical systems, actuation is much more energy-intensive than sensing, and therefore actuators are generally much less numerous than sensors. For example, a human can see objects that are hundreds of meters away, but can only act on objects within a small radius around his or her body — and even then, in a manner severely limited by anatomy, mobility, and strength. Thus, because organisms sense more than they actuate, they contain more internal feedback than forward. Another interpretation is that internal feedback filters sensory information to produce appropriate motor output, performing a sort of dimension reduction.

We remark that this model is based on the standard realization of the state feedback SLS controller, which is internally stable [3]. Internal feedback is a necessary feature in all known alternative realizations of state feedback SLS [71], [72], as well as full control and output feedback SLS controllers. We anticipate that for any local and distributed controller, internal feedback will outnumber feedforward

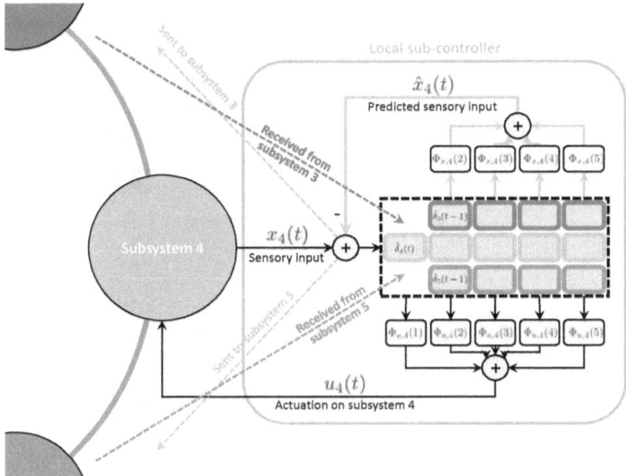

Figure 6.14: Local controller at subsystem 4, assuming subsystem 4 is actuated. The feedforward path is depicted by the black arrows; counterdirectional internal feedback is depicted by the yellow solid arrows, and lateral feedback is depicted by dashed yellow, pink, and green arrows. In this example, we enforce inter-subsystem communication delay; $\hat{\delta}_3$ and $\hat{\delta}_5$ are received from neighboring subsystems with a delay of 1 time step. Note that $\Phi_x(1)$ is not included because for all Φ_x satisfying the feasibility constraint, the $z\Phi_x - I$ (see Figure 2.2) results in $\Phi_x(1)$ being canceled out by I.

paths, though we defer a more thorough exploration of implementation details to future work. Additionally, we remark that in the SLS formulation, any constraint on Φ_x, Φ_u will constrain both the controller and the closed-loop, i.e. local constraints will translate to both local communication and localized behavior. However, we may not always want to constrain the closed-loop — in such cases, we may use alternative techniques, as described in Chapter III. These techniques utilize the same controller structure as standard SLS, so all discussion on memory and feedback still apply to these alternative techniques.

6.6 Speed-Accuracy Trade-Offs

We have shown that state estimation and localization of function require internal feedback to correct for self-generated or predictable movements. In particular, large amounts of internal feedback are required when we take distributed control of a many-sensor, few-actuator system into account. We now show how to efficiently

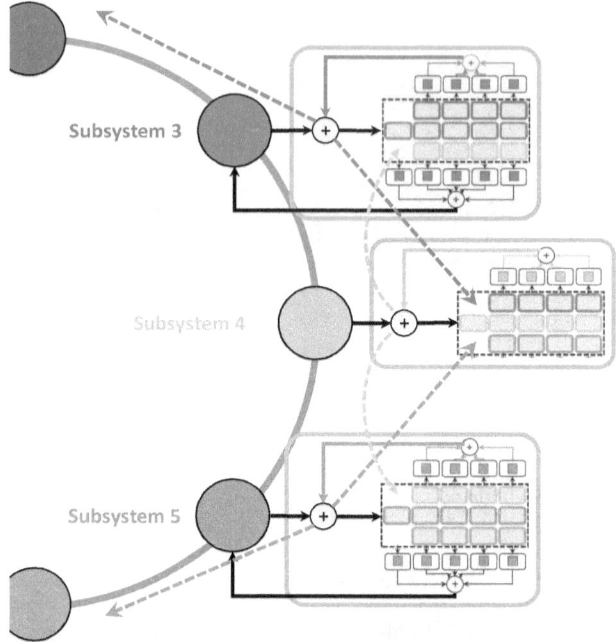

Figure 6.15: Local controllers at subsystems 3, 4, and 5. Each local controller is enclosed in a grey box. Only subsystems 3 and 5 are actuated. Feedforward is denoted by black arrows. Counterdirectional internal feedback are denoted by the solid color arrows, and lateral internal feedback are denoted by dashed color arrows. Not shown are lateral internal feedback from other neighbors of subsystems 3 and 5.

track a moving object with the limitations imposed by neural components and internal time delays using an attention mechanism.

Up to this point, we have assumed that the controller can directly sense the position of the object (perhaps with some delay). In the real world a scene can comprise many objects, which makes it more difficult for a sensorimotor system to localize an object in the scene. However, a moving object, once identified, can be more easily discriminated from a static visual scene. This illustrates the distinction between scene-related tasks (such as object identification) and error-related tasks (such as object tracking), which in the visual cortex is accomplished by the ventral and dorsal streams, respectively.

This distinction also mirrors the separation between bumps and trails in the mountain-biking task studied in [67], allowing us to build on the control architecture in that task. The main difference is that instead of separating into two control loops, we use layering and internal feedback to supplement the control actions of the main control loop.

For simplicity of presentation, we consider a one-dimensional problem (tracking on a line), and use as the metric $\|x\|_\infty$ (worst-case, or adversarial tracking error) rather than $\|x\|_2$ (average-case tracking error). We have some object whose position relative to us, r, is governed by the dynamics

$$r(t + 1) = r(t) + w_r(t) + w_b(t) \tag{6.23}$$

where w_r represents object movement, and w_b represents changes in the background scene. Our limb position p is governed by the dynamics

$$p(t + 1) = p(t) - u(t) \tag{6.24}$$

where $u(t)$ is some limb action. The tracking error $x := r - p$ then obeys the dynamics

$$x(t + 1) = x(t) + w_r(t) + w_b(t) + u(t) \tag{6.25}$$

where the task difficulty is implicitly equal to 1.

We assume that object movement and background changes are bounded: $|w_r(t)| \le \epsilon_r$ and $|w_b(t)| \le \epsilon_b$ for all t. Additionally, we assume that background changes are much slower than object movement: $\epsilon_b \ll \epsilon_r$, i.e.

$$\epsilon_r + \epsilon_b \approx \epsilon_r \tag{6.26}$$

Consider a movable sensor that senses some interval of size β on the continuous line. Information from the sensor must be communicated to the controller via axon bundles, which are subject to speed-accuracy trade-offs — that is, the higher bandwidth a signal, the slower it can be sent. Thus, roughly speaking, axonal communication can be low-bandwidth and fast, or high-bandwidth and slow. We can formalize this as follows for a volume of cortex axons with uniform radius, adapting from [67]:

We first observe that delay T is inversely proportional to axon radius ρ with proportionality constant α

$$T = \frac{\alpha}{\rho} \tag{6.27}$$

Firing rate per axon, ϕ, is proportional to axon radius with proportionality constant β

$$\phi = \beta\rho \qquad (6.28)$$

Cross sectional area s is related to axon radius ρ and the number of axons in the nerve n via

$$s = n\pi\rho^2 \qquad (6.29)$$

And finally, signaling rate R of the entire nerve, which is related to the resolution of information sent about the sensed interval, is represented by

$$R = n\phi \qquad (6.30)$$

These equations can be combined to obtain the speed-accuracy trade-off

$$R = \lambda T, \quad \lambda = \frac{s\beta}{\pi\alpha} \qquad (6.31)$$

The constant λ is proportional to s, the cross-sectional area; for projections of fixed length, this represents the spatial and metabolic cost to build and maintain the axons. In general, given some fixed cortical volume, we can either build few thick axons, which will have low delay but low information rate, or we can build many thin axons, which will have high information rate but high delay.

We implement this speed-accuracy trade-off using a static, memoryless quantizer Q with uniform partition, followed by a communication delay, as shown on the top in Figure 6.16. This choice of quantizer does not add to the cost, since it recovers the optimal cost over all possible quantizers [73]. The controller can move the sensor around; the interval sensed by the sensor remains constant, but the controller can choose where the interval lies. Assume the initial position of the object is known — we can select an initial sensor location and β appropriately such that $r(t)$ always falls within the sensed interval. In this case, the best possible tracking error for any delay T is

$$\epsilon_r T + \frac{\beta}{2\lambda T} \qquad (6.32)$$

The first term represents error from delay, object movement, and drift. In the time taken for information to reach the controller, the most adversarial action possible by the object and background would contribute a tracking error of $(\epsilon_r + \epsilon_b)T$; we apply simplification (6.26) to obtain $\epsilon_r T$. The second term represents quantization error.

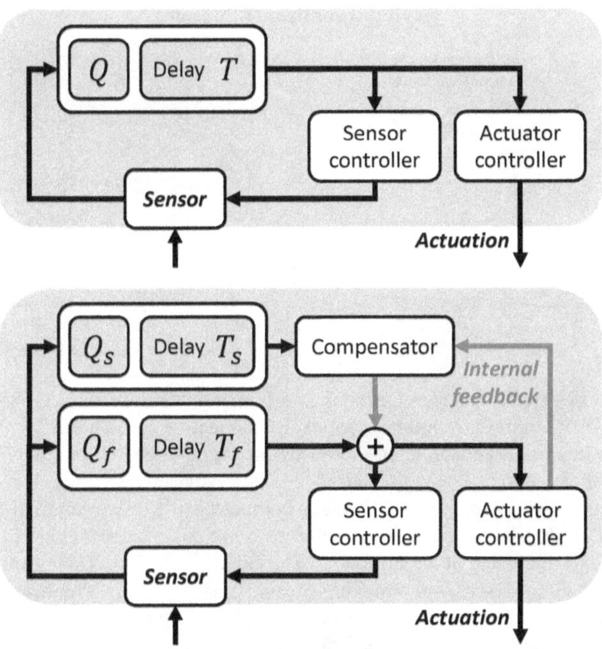

Figure 6.16: Optimal control model of attention, with moveable sensor. **(Top)** Model with one communication path, in which information is quantized by quantizer Q and conveyed to the controller with delay T. **(Bottom)** Model with two communication paths, and two separate quantizers Q_s, Q_f and respective delays. This model can be considered lateral (e.g V1-V1) or counterdirectional (V2-V1) internal feedback (pink) between the two controller paths.

For an interval of size β divided into N uniform sub-intervals, the worst-case error is $\frac{\beta}{N}$; we then use the fact that $N = 2^R = 2^{\lambda T}$.

This is achieved by the controller depicted on the top in Figure 6.16. The cost, as a function of T, is plotted in the left panel Figure 6.17 with the label "No Internal Feedback" (where $T = T_s$). Here, the speed-accuracy trade-off is implicit. Very low values of T correspond to very low signaling rates — the controller does not receive enough information to act accurately, so performance is poor. The opposite problem occurs at very high values of T; though the information is high-resolution, the time elapsed between information and action is too long, leading to poor performance. The best performance occurs between these two extremes [67].

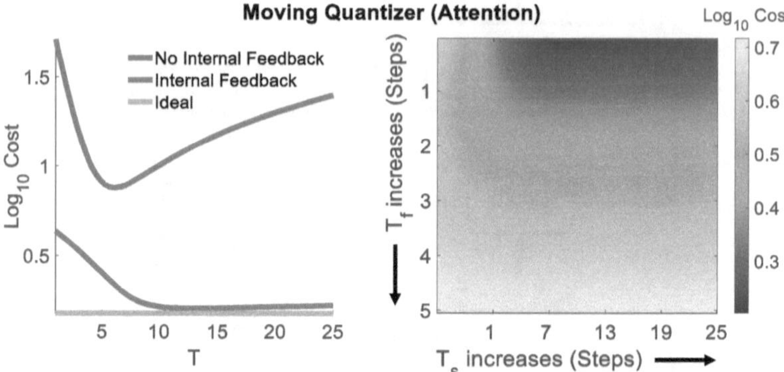

Figure 6.17: **(Left)** Internal feedback and layering achieve superior performance when sensor-controller communications are subject to speed-accuracy trade-offs. The "No Internal Feedback" controller uses one layer, while the "Internal Feedback" controller uses two layers, with internal feedback between the layers. The two-layer case consists of a fast forward pathway compensated by slow internal feedback, which takes slow background changes into account; this achieves better performance (lower cost) than the case without internal feedback. The "Ideal" controller, where the sensor directly senses the moving object, is also shown. The layered system with internal feedback achieves performance close to the ideal. Task difficulty is $\alpha = 1$. T represents delay. For the "No Internal Feedback" controller, it represents the delay of the single layer; for the "Internal Feedback" controller, it represents the delay of the slow layer, i.e. $T = T_s$. The delay of the fast layer is held constant. **(Right)** Performance (log cost) of the two-layer controller with internal feedback as delays of both layers are varied. Notice that performance is generally good when T_f is low and T_s is sufficiently high.

We can improve this performance by nearly an order of magnitude by adding an additional communication pathway and the requisite internal feedback. We now have two communication paths from the sensor, each with its own quantizer and delay block. The slower communication path uses quantizer Q_s with delay T_s, while the faster path uses Q_f with delay T_f. To further facilitate speed in the fast path, we allow it to send only a subset of information from the sensor (i.e. only send information about a small part of the sensed scene). Mathematically, let the fast path send information about an interval of size β_f, with $\beta_f < \beta$, and let this smaller sub-interval be contained within the sensor interval. This sub-interval is an implementation of *attention*. The fast path is the main actuation path, while the slower path provides compensatory signals via internal feedback; this is shown on the bottom in Figure 6.16. In this case, the best possible cost is

$$\epsilon_r T_f + \frac{\beta_f + E_s}{2^{\lambda T_f}}$$
$$E_s = \epsilon_b T_s + \frac{\beta}{2^{\lambda T_s}}$$

(6.33)

The first term represents error from delay and object movement, similar to (6.32). The second term represents a combination of quantization error from the fast communication pathway (β_f) and performance error of the slow pathway (E_s), which informs the fast pathway of where to place the sub-interval. Notice that E_s takes the same form as (6.32).

The cost, as a function of T_s, is plotted in the left panel of Figure 6.17 with the label "Internal Feedback". In this plot, we assume T_f to be its smallest possible value: one unit of delay. We see that using two quantizers in combination with internal feedback is superior to using one quantizer. We remark that this only holds when the two quantizers are *different*; if we simply use two quantizers with the same interval, bit rate, and delay, no such performance boost occurs. In general, holding T_s constant and decreasing T_f improves performance, as shown in the right panel of Figure 6.17.

Functionally, the inclusion of a faster communication pathway allows action to be taken in a more timely manner than in the single-pathway case. Unlike in the single-pathway case, we are not encumbered by issues of low-resolution information; the slower communication pathway corrects the fast pathway through internal feedback. Here, as in previous examples, the internal feedback carries signals correcting for self-generated and slow, predictable changes. Overall, despite speed-accuracy trade-offs in communication, the system achieves fast and accurate behavior with the help of internal feedback, under reasonable assumptions about the dynamics of the scene and environment.

6.7 Discussion and Interpretation of Results

We analyzed a set of minimal control models to explore internal feedback in a perception-action loop with time delays and limited communication bandwidth. We showed how previously internal feedback, which is unbiquitous in brains and poorly understood, facilitates state estimation and localization of function, and how attention facilitates motor performance. This is a step toward an end-to-end model of sensorimotor processing in neural systems.

The mathematical framework for biological control explored here can be applied across a wide range of experimental systems. The simple models were meant to provide an intuitive understanding of the control strategies for handling internal time delays and limited signaling bandwidth. The framework can be elaborated to make more specific predictions for more complex biological systems.

Task-oriented whole-system frameworks reveal new roles for internal feedback
There may be other functions for internal feedback in addition to compensating for time delays and limited cortical communication bandwidth that have arisen in parallel with the development of methods for increasingly high-throughput and high-resolution biological measurements that probe internal feedback and internal dynamics, essential features of cortical computation and control. Additional functions that have been suggested are computation through dynamics [63], deploying recurrent networks [64], performing Bayesian inference [65], [66], generating predictive codes [61], [62], and many others.

These frameworks emphasize prediction but are largely confined to the context of either sensory processing or motor processing separately and do not explicitly model closed loop task performance. Our analysis considers sensorimotor control from an ethological perspective, which is important because the ultimate selection pressure on sensory processing is to support optimal actions for ensuring survival [74].

Our framework is is consistent with recent neural recordings from the cortex showing that motor signals and the influence of past and current actions account for substantial cortical activity, previously considered spontaneous, background or noise [56]–[59], [75]. These internal feedback signals carry information about how actions propagate through the body and its environment, ameliorating communication limitations that affect both plans and future actions.

The predictions of these models can be used to interpret how ablation, suppression, or delay of counterdirectional internal feedback would degrade performance in many visuomotor tasks. The performance gap should be more pronounced in tasks that involve quickly changing conditions. Similar interventions that disrupt lateral feedback interactions within motor cortex should degrade performance in tasks where body parts are highly mechanically coupled — such as hands and shoulders — and less so for tasks where body parts are loosely mechanically coupled, such as speech and walking.

Our theory predicts that the fastest neurons from V1 to V2 or V1 to MT (including

Meynert cells) should be the most highly activated when there are unpredictable changes in the visual scene. Similarly, we attribute autonomous dynamics in M1 to the communication of predicted actions laterally and counterdirectionally. This leads to the prediction that unexpected perturbations during a motor task should lead subpopulations of cells in M1 (including Betz cells) to transmit rapid change-related signals in contrast to the smooth responses that accompany unperturbed control.

Classical optimal control models neglect physiological limitations

Optimal control theory is a general framework for sensorimotor modeling. Given a mathematical description of a system and some task specification, the optimal controller provably gives the best possible performance. However, these proofs assume that the components are fast and accurate, with instantaneous communication and control circuits implemented with fast and accurate electronics. Using these components, a single sense-compute-actuate loop is generally sufficient to achieve optimal behavior.

Applying control theory to model physiological circuitry requires a distinction between *behavior* and *implementation*. The same optimal performance may be achieved through a number of different implementations in the underlying circuitry. Although traditional control theory excels as a model of sensorimotor behavior, it does not incorporate the component-level constraints that are prevalent in biology; as a consequence, the ways that traditional control theory models are implemented may not be directly relevant to biological control.

Recent advances have extended traditional control theory to allow distributed control and incorporation of component-level constraints [3], [67], [69], [70]. We build on this body of work to describe how constrained components affect the implementation of an optimal distributed biological controller. In particular, we show how and why internal feedback arises in controllers whose components exhibit the speed-accuracy trade-offs found in brains.

Fast long-range association fibers in the cortex are metabolically and developmentally expensive, have low bandwidth (compared to slower fibers with higher bandwidth), require constant maintenance and repair and are limited in number. Internal feedback from the motor cortex to earlier sensory areas can regulate the use of these pathways by suppressing self-generated and other predictable signals, freeing the fast pathways to selectively transmit the unpredicted changes needed by the motor system to make make fast decisions. This virtualizes the behavior of the control system to produce actions that are both fast and accurate despite internal time delays and limited communication bandwidth.

Evidence for the suppression of self-generated sensory signals by corollary discharge signals

Efference copies of motor signals are ubiquitous throughout brains and serve several functions. They are used for example as the input to forward models to predict the consequences of motor commands. Another use of efference copies is for corrolary discharges that suppresses predictable self-generated sensory signals. these help to distinguish sensory signals arising in the environment from those that are self-generated, which are often much larger and can interfere with motor tasks, as occurs when there is a long time delay between a microphone and a loudspeaker. Fast suppressive internal feedback signals originate before motor commands are executed and target sensory pathways before self-generated signals can reach higher levels of processing [76]

Perhaps the best understood neural system that suppresses predictable self-generated sensory signals is found in electric fish, which generate electric fields for navigation and communication [77]. The electrosensitive lateral-line lobe (ELL), a cerebellum-like structure, receives both a corollary discharge of the generated electric field and sensory input from electroreceptors on the body of the fish. These two signals are subtracted in the ELL to detect externally generated electric fields. Suppression is learned using anti-Hebbian synaptic plasticity, in which the coincidences of incoming spikes and outgoing spikes lead to a decrease in synaptic strength. A similar arrangement is found in the dorsal cochlear nuclei of mammals, which receives corollary discharge signals from brainstem nuclei associated with vocalization and respiration as well as proprioceptive input from body movement [77].

Biophysical speed-accuracy trade-offs drive internal feedback

Biological control systems do not have components that are both fast and accurate. Spiking neurons, though fast relative to other biological signaling mechanisms, are many orders of magnitude slower than electronics and face severe speed-accuracy trade-offs that constrain communication and control. However, by cleverly combining components with different speed-accuracy trade-offs and using internal feedback as demonstrated above, brains are able to perform survival-critical sensorimotor tasks with speed and accuracy.

Neurons have biophysical constraints that lead to speed-accuracy trade-offs. For example, some neurons can rapidly convey a few bits of information, and others can slowly convey many bits of information, but neurons that rapidly convey many bits of information are expensive and correspondingly rare. Speed-accuracy trade-offs include spike averaging versus spike timing and the spatial trade-off between the number of neurons (information rate) and their axonal diameter (conduction speed) in nerve bundles [67], [78]. These trade-offs have consequences for the performance of sensorimotor systems that we can study in our control models.

Brains contain highly diverse populations of neurons. For example, the range of neural conduction speeds in humans spans several orders of magnitude [78]. In sensorimotor systems, these diverse neurons are multiplexed in a task-specific way that approximates the performance of a single-loop system composed of ideal (e.g. fast *and* accurate) components [67]. The fastest components are used in the feed-forward loop, sending information from sensing areas toward motor areas. Internal feedback compensates for accuracy by filtering out slow-changing, predictable, or task-irrelevant stimuli, such that the fewest possible bits need to travel along the fastest possible neurons. From an evolutionary perspective, once a system can achieve fast responses, additional layers of control can be added to achieve more accurate and flexible behavior without sacrificing performance.

The reason why internal feedback is not needed in most engineered control systems is that internal time delays are negligible. But in biological systems, even the fastest neurons used in the feedforward loop give rise to significant time delays. This is why it is essential to include delays in our control-based analyses of the forward loop in models of control in brains (Fig. 6.3).

Fast forward conduction is key to successful sensorimotor task performance
In cortex, the fastest, largest and most striking neurons are the large pyramidal cells: Meynert cells in primary visual cortex carry signals from rapidly moving-object; giant Betz cells in motor cortex that project to the spinal cord are responsible for rapid responses to perturbations from planned movements; and giant Von Economo cells in the prefrontal cortex (anterior cingulate and fronto-insular areas) that project subcortically are involved in the regulation of emotional and cognitive behavior. [79]–[82].

The visual stream diverges into the dorsal and ventral streams, which are responsible for object motion and object identity, respectively. In natural scenes, object locations may change quickly, but object identities are change relatively slowly; a mouse may move around rapidly, but its status as a prey hunted by a barn owl does not change. Thus, speed is crucial for the dorsal stream, but not the ventral stream.

This difference has physiological consequences in our minimal model of attention that could explain differences between cortical projections in the two streams: the giant Meynert Cells that project from V1 to MT (an object motion area in the dorsal stream; see Figure 6.1), but there are no equivalently large cells projecting from V1 to inferotemporal cortex (an object identity area in the ventral stream). Reaching tasks could test the predictions of our control model for rapidly and unpredictably moving objects on a fixed background compared with predictably moving objects on nonstationary backgrounds.

Neurons in MT respond selectively to the direction of moving objects and provide signals that are used by the oculomotor system for the smooth pursuit of moving objects [83].There are several visual pathways from the retina to the cortex for tracking moving stimuli. In addition to the cortical pathway that projects from V1 to area MT in Figure 6.1, the retina also projects to the pulvinar, another thalamic relay to extrastriate areas of the cortex [84]. These two pathways could implement the optical control model in the bottom panel of Figure 6.16, where the fast, direct pathway is from the pulvinar and the delayed, indirect pathway from V1 corresponds to the slower pathway.

Internal feedback facilitates fast feedforward signals in visual cortex

In recent years, large-scale recordings from visual cortex have uncovered non-visual signals that challenge the traditional single-loop view of sensorimotor control. In the traditional view, visuomotor processing consists of a series of successive transformations from stimulus to response, with each cortical area along the way tuned to some aspect of stimulus space [85]. However, although V1 does respond to visual stimuli, the activity of these neurons is also characterized by motor-based internal feedback and task/attention-related modulatory internal feedback [54], [58], [59], [74]. Our attention model implements the observed enhancement of responses in neurons that selectively respond to an attended stimulus through internal feedback [86].

The number of projections from V1 to V2 is roughly the same as number of neurons, of similar conduction speed, that project from V2 to V1 [50]–[52]. However, these neurons are very different in morphological and molecular characteristics: the neurons that project feedforward from V1 to V2 primarily activate AMPA receptors, while the feedback neurons that project from V2 to V1 have a strong NMDA receptor component and terminate almost exclusively on excitatory pyramidal neurons [87], [88]. Both of these receptors are activated by glutamate, but AMPA-mediated currents are fast, lasting only a few ms, while NMDA-mediated currents can linger in the postsynaptic neurons for hundreds of ms [89]. This feedback could be relevant for top-down signaling to shape and control perception during actions. Because NMDA receptors trigger synaptic plasticity, the feedback could also be important for learning how to suppress self-generated sensory signals as well as perceptual learning.

Pharmacologically blocking NMDA receptors in visual cortex disrupts figure-ground discrimination; that is, a loss in capacity to contextually interpret the visual scene [90]. In the context of our theory and minimal model of attention, internal feedback from V2 informs V1 of predictable elements in future stimuli.

Since the visual space cannot be sampled losslessly, these feedback signals could be helping V1 suppress predictable features, making the the unpredictable features more salient [87].

Internal feedback facilitates localization of function in motor cortex
Primary motor cortex (M1) is dominated by its own past activity rather than static representations [55]. In the context of the state estimation problem we considered in Fig. 6.8, these dynamics in motor cortex are driven neither by motor representations nor by pattern generation, but by predictions of the consequences of self-action through local internal feedback, which need to be sent throughout the body.

By the same principle, the localization of function within motor cortex that we considered in Figure 6.10 reconciles the conventional view of homuncular organization with, for example, the body-related signals found in putatively hand-related parts of motor cortex, as well as contralateral hand signals [56], [57], [91]. As with motor signals in visual cortex, these broad body movement signals in motor cortex are crucial for identifying predictable consequences of motor signals from other body movements and separating them from unpredictable signals of critical importance for rapidly controlling localized body parts. This provides each body part with the context it needs to compensate for the movement of other body parts.

In our analysis, we assumed the existence of a distinct motor cortex that generates motor commands and a visual cortex that interprets visual scenes and is essentially an extension of the retina. With these assumptions, we consider where internal feedback should exist. Is it possible for the entire estimator to be implemented in the visual cortex? Since the estimator uses predictions of future actions, the estimator requires at least some input from motor cortex.

Is it possible for the entire estimator to be implemented in motor cortex? The dynamical structure of responses in the motor cortex is compatible with a predictive and delay-compensating function of exactly the kind our model suggests. However, our model also shows that the information transmitted from sensor to motor cortex depends on the estimator. Thus, for motor cortex to carry out this entire function without anatomically counterdirectional internal feedback, it would need to contain all of visual cortex. The counterdirectional internal feedback then solves this problem with a loop between the visual cortex and the motor cortex

In addition to feedback loops between the motor cortex and other cortical areas, there are also loops with the cerebellum and the basal ganglia. These provide additional information about sensory predictions and sequences of future actions, respectively. Regions of the motor cortex that interdigitate between projections to body parts have recently been identified that are associated with stimulation-evoked complex actions and connectivity to internal organs such as the adrenal medulla that are associated with goals [92]. These regions may be responsible for integrating skeletal body movements with visceral states and goals.

Learning on internal feedback pathways fine-tunes performance

Internal feedback pathways carry attentional signals that activate slow NMDA receptors, which in turn regulate the strengths of synapses [93]. We have shown that internal feedback pathways are needed for ignoring self-generated and other predictable signals. Early in brain development, activation of NMDA receptors in primary visual cortex before the first visuomotor experience is needed to suppress predictable feedback and the the selection of unpredictable stimuli [90]. Knocking out these NMDA receptors impairs ongoing visuomotor skill learning later in life, which may compensate for body growth and body weight changes. Learning to to reduce self-generated sensory prediction error can be implemented locally through same internal feedback system that broadcasts motor predictions.

Reinforcement learning governed by circuits in the basal ganglia may also benefit from the internal feedback pathways in the cortex. Transient dopamine release, which carriers reward prediction error, does not specify which sensory inputs were responsible for the reward, which in part is why it is a much slower form of learning.

Attentional internal feedback in the cortex automatically selects and represents the currently most salient information to guide motor actions. Attentional information projects to the striatum and makes it easier for the basal ganglia to associate the causally relevant sensory inputs with reward signals [74].

We have proposed that prediction is an essential aspect of performance in visuomotor tasks where fast and accurate responses are needed. Prediction is useful in the model because it enables compressed representations of the current state, which can then be transmitted more quickly across the nervous system because of the speed-accuracy tradeoff discussed above. Extending this principle, high performance can be achieved by simpler representations of tasks, which in turn allow faster responses.

After a flexible but slow learning system has successfully mastered a task, it could then be 'loaded' onto a simpler, more rigid, and faster subcortical system. Internal feedback can facilitate this transfer. We have focused here on the fast pathways represented by large axons in visuomotor cortex, but similar variation in timescales of conduction can be found throughout the sensorimotor system. We, therefore propose that during the acquisition of fast and accurate motor skills, control would shift from slower learning systems in the cortex to less flexible subcortical parts of the motor system.

Attention has been studied primarily in the context of sensory processing. The importance of attentional signals for reducing time delays in making motor decisions adds a new direction for future experimental studies. Attention is linked to conscious awareness and rides atop the global representation of the body throughout the cortex. This makes internal feedback a candidate feature of the nervous system that helps explain the sense of unity that we experience, which would otherwise be difficult to achieve within a balkanized control architecture built on body parts.

6.8 Conclusions and Future Work

This chapter focuses on incorporating delay, functional localization, local communication, and bandwidth limitations into sensorimotor control models. While system level synthesis provides a general theoretical framework which we can use to study delay and local communication in controllers, we lack a similar framework for bandwidth limitations in controllers. One direction of future work is to produce such a framework. Furthermore, as alluded to earlier, any controller that is expressed in terms of transfer functions can be realized and implemented in a variety of ways; in this chapter, we have taken the most natural realization and implementation, but other implementations remain unexplored. For instance, instead of an explicit memory, memory may be implicitly contained in delayed wires. Clarifying the space of realizations and implementations is a critical step toward the ultimate goal of this work, which is to elucidate how neurons (with all their signaling limitations) coordinate to implement optimal controllers in organisms.

Chapter 7

A LAYERED MODEL OF DROSOPHILA LOCOMOTION

This chapter is based on a manuscript that is presently under preparation. The current authors are Lili Karashchuk (co-first author), myself (co-first author), John C. Tuthill, and Bing W. Brunton. I contributed to framework formulation and manuscript writing, and was solely responsible for all work pertaining to the dynamics and optimal controller. All presented results are preliminary. Code needed to reproduce model-generated walking and walking data timeseries (as presented in this Chapter) are provided as supplementary files to the book.

Overview: We present a novel layered model of *Drosophila melanogaster* that generates realistic and robust multi-legged locomotion. The model is composed of three functional layers, or modules: an inter-leg coordinator, a data-driven neural network that generates realistic per-leg kinematics, and an optimal controller to enact the desired kinematics in a physical environment, subject to motor delays. We use the model to generate walking at various forward and turning speeds and find that the resulting angle trajectories are highly similar recorded trajectories from real flies. The modular nature of the model lends itself to generalization; we demonstrate that the model maintains walking behavior in the presence of slip-like perturbations, despite not being trained for such perturbations. Additionally, we leverage the model to provide insight on fundamental physiological limits — allowable ranges of motor delays that preserve robust fly walking.

7.1 Introduction

Legged locomotion requires coordination between many elements: central pattern generators and inter-leg coupling, control of leg poses via muscles, and integration of proprioceptive feedback [94]. The ultimate goal of the study of legged locomotion is to produce analyses and models that unite physiology, physical dynamics, and behaviors and uncover the design principles behind locomotion. To this end, many elements of legged locomotion have been studied in various combinations for hexapods, predominantly flies and cockroaches. Inter-leg coupling is studied via analysis of kinematic data in [95], [96], and [97]. Locomotion patterns are studied in the context of energetic cost in [98].

To produce models of walking behavior, one approach is to tune parameterized networks of coupled oscillators. [99] and [100] use networks of coupled oscillators and motoneurons, tuned to recreate gaits from data. These works model each leg of the organism as a single oscillator, and do not include details on individual joints of legs — however, they leverage data to produce realistic oscillatory trajectories. Another approach is to focus on physical details: [101] and [102] introduce a robotic platform, while [103] introduces a physics-engine based platform. [104] uses a decentralized reactive controller to recreate hexapod gaits. These models are also composed of tuned oscillators and controllers: the resultant walking behaviors, though dynamically valid, deviate substantially from those of real organisms in terms of joint kinematics. A third popular approach is to use data-free learning and optimization to generate learning-type behaviors from scratch [105], [106]. These approaches produce walking with varying degrees of realism, but require clever selection of objectives and constraints, and are computationally expensive — these also do not consider physiological details (e.g. neuronal constraints) beyond biomechanics.

In short, existing approaches typically lack at least one of the following: physiological considerations, physical dynamics, or realistic behavior. We aim to address this in this chapter. We use a layered model that produces realistic joint-level kinematics and incorporates physiological motor delays. This end-to-end model that combines features of many previous studies and models, including coupled oscillators, data-driven neural networks, optimal controllers, and link-and-joint leg dynamics. The dynamics are controlled by the optimal control layer, which receives proprioceptive feedback and maintain realistic trajectories while rejecting external perturbations and accommodating motor delay. Realistic joint trajectories are generated by a neural network, which is trained on kinematic data collected from hundreds of bouts of Drosophila walking [107]. The top-most layer consists of coupled oscillators, which coordinate walking phases between legs.

A key feature of this model is its modularity: together, the layers of the model achieve more than the sum of its parts. A particularly important piece is the combination of the neural network and the optimal controller; a pure neural network is physiologically uninformative and does not generalize well with external perturbations, while a pure optimal controller cannot produce realistic kinematics; however, the combination of the two produces realistic kinematics *and* is robust to external perturbations. In other words, the incorporation of the optimal controller allows the data-driven model to generalize to new scenarios as well as interface with physical dynamics. Additionally, in our model, a novel controller formulation is used to incorporate motor delays, which are key physiological factors in biological locomotion.

7.2 End-to-End Learning and Control Model

We propose an end-to-end model of *Drosophila* walking, from single leg joint dynamics to inter-leg coordination. The efficacy of the model is demonstrated by comparing model-generated simulations with real data. The model employs three functional layers, which interface with a dynamics model, shown in Figure 7.1. The three layers are the optimal controller, the trajectory generator, and the phase coordinator. Each individual leg is governed by its own dynamics, optimal controller, and trajectory generator, while inter-leg coordination is accomplished by the phase coordinator.

The use of this layered architecture allows us to unify different aspects of walking; individual leg dynamics, individual leg kinematics, and inter-leg coupling. Each layer provides an abstraction for the layer above it, such that the different aspects of walking can be modularly integrated.

Dynamics and control

Leg dynamics for each leg are derived from ball-and-joint models. We begin with a link-and-joint model of the fly leg, as depicted in Figure 7.1. For simplicity, we only model joints that are crucial to natural leg movements during walking and turning. For instance, varying femur rotation is important to the movements of the middle legs, but the front legs exhibit near-constant femur rotation; thus, a femur rotation joint is included for the middle and hind legs only. The joints included for each leg is shown in Table 7.1.

Table 7.1: Joints included for leg models

Joint	Front legs	Middle legs	Hind legs
Body-coxa flexion	✓		
Coxa rotation	✓		
Coxa-femur flexion	✓	✓	✓
Femur rotation		✓	✓
Femur-tibia flexion	✓	✓	✓

We write the Denavit-Hartenberg (DH) table of the leg model, and use this to

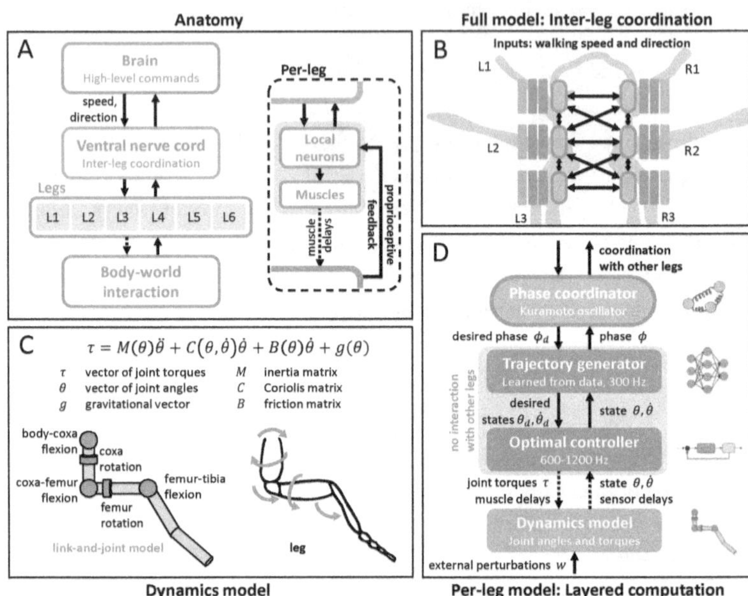

Figure 7.1: Summary of layered locomotion model and relation to anatomy. **(A)** Anatomy involved in walking. The brain sends high-level commands (e.g. walking speed and direction) to the ventral nerve cord (VNC), which coordinates across legs. Each leg's neurons and muscles take inputs from the VNC and acts on the environment; this body-world interaction and its effect for leg joints, joint angular velocities, torques, etc. are reported back to local circuits on the leg via proprioceptive feedback. **(B)** Full model: per-leg models are coupled through their phase coordinators (blue ovals). **(C)** Per-leg dynamics model, derived from link-and-joint models and Euler-Lagrange equations. **(D)** Layered per-leg model. Body-world interactions are modeled through a dynamics model. We assume proprioceptive feedback provides information on joint angles and angular velocities. Each leg contains an optimal controller which interfaces with dynamics, and a trajectory generator which generates realistic gaits. The trajectory generator interfaces with the phase coordinator, a Kuramoto oscillator which induces inter-leg coupling. The trajectory generator and optimal controller mimic local circuits and do not interact with other legs.

systematically derive the Euler-Langrange matrix equations of motion

$$\tau = M(\theta)\ddot{\theta} + C(\theta, \dot{\theta})\dot{\theta} + B(\theta)\dot{\theta} + g(\theta) \tag{7.1}$$

where τ is the vector of joint torques; θ, $\dot{\theta}$, and $\ddot{\theta}$ are vectors of joint angles, angular velocity, and angular acceleration; M, C, B, are the inertia, Coriolis, and friction matrices; and g is the gravity vector.

We then define state $q = \begin{bmatrix} q_1 \\ q_2 \end{bmatrix} = \begin{bmatrix} \theta \\ \dot{\theta} \end{bmatrix}$ and input τ, and rearrange (7.1) into the form $\dot{q} = F(q, \tau)$, i.e.

$$\begin{bmatrix} \dot{q}_1 \\ \dot{q}_2 \end{bmatrix} = \begin{bmatrix} q_2 \\ -M(q_1)^{-1}\left(C(q_1, q_2)q_2 + B(q_1)q_2 + g(q_1)\right) \end{bmatrix} + \begin{bmatrix} 0 \\ M(q_1)^{-1} \end{bmatrix}\tau \tag{7.2}$$

We then choose equilibrium values $\bar{q}$ and $\bar{\tau}$, such that $F(\bar{q}, \bar{\tau}) = 0$. $\bar{q}_1$ is chosen based on the average joint angles for each leg, from the data. Then, $\bar{q}_2 = 0$ and $\bar{\tau} = g(\bar{q}_1)$ gives the desired equilibrium. We linearized about this equilibrium point, which leads to the following equations

$$\dot{x} = A_c x + B_c u \tag{7.3}$$

where A and B are the Jacobians with respect to q and τ, respectively, i.e. $A = \frac{\partial F}{\partial q}(\bar{q}, \bar{\tau})$, $B = \frac{\partial F}{\partial \tau}(\bar{q}, \bar{\tau})$; and $x := q - \bar{q}$ and $u := \tau - \bar{\tau}$. In our code, we use the SymPyBotics toolbox [108] to obtain symbolic equations for the quantities in (7.1), then numerically compute Jacobian values.

Next, we discretize the system using some sampling interval T, which is typically chosen to be an integer multiple of sampling interval from the data ($T = 1/300$). In our simulations, we use $T = 1/600$. The discretized dynamics are written as

$$x(t + T) = Ax(t) + Bu(t) \tag{7.4}$$

where $A = I + A_c T$ and $B = B_c T$.

Finally, we perform a coordinate shift to error dynamics. This allows us to apply standard control techniques for trajectory tracking. Define tracking error $y = q - q_d$. This error obeys the following dynamics

$$y(t + T) = Ay(t) + Bu(t) + w(t) + w_{traj}(t) \tag{7.5a}$$

$$w_{traj}(t) = A(q_d(t) - \bar{q}) + \bar{q} - q_d(t + 1) \tag{7.5b}$$

where w is external perturbation, and w_{traj} represents the effect of constantly changing trajectories. Error y is of size n_y (8 for front legs, and 6 for the other legs), and input u is of size $n_u := n_y/2$.

To include actuation and sensor delay, we make use of augmented state formulations as introduced in Chapter VI. Let the actuation delay be d_{act} steps, and let the sensor delay be d_{sense} steps. Our augmented state vector $z(t)$ is written as

$$z(t) = \begin{bmatrix} y(t) \\ f(t) \\ a(t) \\ s(t) \\ g(t) \end{bmatrix} \tag{7.6}$$

where y is the error from (7.5); f is size $d_{act} * n_y$, containing predicted future errors up to d_{act} time steps in the future; a is size $d_{act} * n_u$, containing actuation signals from up to d_{act} steps ago; s is size $d_{sense} * n_y$, containing sensing signals from up to d_{sense} steps ago; and g is size $d_{act} * n_y$, containing information about future trajectory effects w_{traj} up to d_{act} time steps in the future.

We write the overall system in the form of

$$z(t + 1) = Fz(t) + Gu(t) + w_{aug}(t) \tag{7.7a}$$

$$r(t) = Hz(t) \tag{7.7b}$$

where F, G, and H are formulated using A and B according to techniques from Chapter VI, and $w_{aug}(t)$ contains the perturbations from (7.5), appropriately rearranged and zero-padded.

Note that u represents the actuation signal, which must be delayed for some amount of timesteps (via a) before affecting the system; similarly, the sensor signal is delayed for several timesteps (via s) before reaching the controller via r.

To achieve effective trajectory tracking, we seek a control law under which y remains small. This can be achieved using the Linear Quadratic Gaussian controller, a standard technique. The controller is governed by

$$\hat{z}(t + 1) = F\hat{z}(t) + Gu(t) + L(r(t) - H\hat{z}(t)) \tag{7.8a}$$

$$u(t) = K\hat{z}(t) \tag{7.8b}$$

where $\hat{z}$ is the estimate of the state, estimated via a steady-state Kalman filter. L and K are the optimal observer and controller matrices, respectively, synthesized via discrete algebraic Ricatti equations.

Overall, the dynamics equations relate the state (angle and angular velocity of each joint) of the leg model with the actuation (muscle-generated torque on each joint). At regular intervals, the controller receives a time series of desired state trajectories from the trajectory generator layer. The controller then produces the necessary torques to track (i.e. recreate) this trajectory. External perturbations, when present, enter through the dynamics and affect the states; the controller then senses the state and reacts to them accordingly.

We remark that the walking and perturbation-rejecting capabilities of the overall model are not contingent upon any specific dynamics or controller formulation; any controller that adequately tracks the trajectory generator will suffice.

Trajectory generator

The trajectory generator is a per-leg neural network, trained on biological data [107]. It takes the current leg phase, joint angles, and joint angular velocities as input, and outputs the phase, joint angles and angular velocities for the next time step. It can be recursively called to to generate a time series of phase, joint angles, and angular velocities for each leg.

The trajectory generator periodically receives proprioceptive information from the controller on the true state (joint angles and angular velocities) of the leg. It also receives information on the desired phase from the phase coordinator. Using this information, it generates the desired angle and angular velocity trajectories for some future time interval, and sends this back to the controller; it also generates phase values, which are sent to the phase coordinator.

In the absence of external perturbations, the trajectory generator generates realistic walking; it is the key to the overall model's biologically accurate kinematics. When external perturbations are present, their effects are largely mitigated by the controller layer; the trajectory generator itself generally receives low magnitudes of perturbation, and maintains relatively accurate kinematics. We remark that all data used to train the trajectory generator come from experimental conditions with no external perturbations.

Phase coordinator

So far, we have described a per-leg dynamical model, per-leg controllers, and per-leg neural networks. To coordinate the action of these six legs, we utilize a Kuramoto oscillator as a phase coordinator. Given some walking speed and current leg phases (produced by the trajectory generators), the phase coordinator produces the desired leg phases as output. The desired leg phases are used by the trajectory generators of each leg to produce future trajectories.

7.3   Realistic Model-Generated Walking

Various metrics and visualizations of simulated vs. real walking are shown in Figure 7.2; these metrics are explained in the following subsections. All simulated trajectories were produced with motor delays of 30 ms, which is consistent with experimentally obtained values of motor delay [109]. Though our framework allows the incorporation of sensory delays as well, we opt to focus on motor delay.

Example time series

We include example time series of simulated walking vs. real walking, as plots of angles and angular velocities versus time for specific joints on a specific leg. This is shown in Figure 7.2, panels A and C. Simulated values resemble real values; real values are comparatively noisier, but the general pseudo-triangular shape, as well as amplitude and frequency of angular oscillations, are similar. The purpose of these visualizations is to demonstrate *qualitative* rather than quantitative resemblance between simulated and real walking.

Angles and angular velocities versus phase

We compare simulated walking and real walking data by comparing the angles and angular velocities of the two. The naive approach is to compare between time series — this can be misleading, since even in data, the time series for two flies walking in the same direction at the same speed may differ significantly due to phase misalignment. The appropriate comparison is to compare angles and angular velocities as functions of phase. Here, we make the distinction between *generated phase*, the per-leg phases produced by the phase coordinator of the model, and *computed phase*, which can be computed for each joint from time series data. For real walking data, we do not have access to the generated phase; thus, comparisons must be made between computed phases.

In panels B and D of Figure 7.2, we plot the angle and angular velocity vs. phase for representative joints (e.g. femur-tibia flexion) for real vs. simulated walking for single bouts, at varying forward-walking speeds (4 mm/s, 8 mm/s, 12 mm/s, 16 mm/s). The real and simulated data are highly similar.

The aggregate differences between simulated and real data over a range of forward-walking, backward-walking, turning, and side-stepping speeds are shown in panel E. The dotted line indicates average in-sample differences between different bouts of real walking — we see that the average difference between the model and data is comparable to the in-sample differences in the data itself. We note that the errors for angular velocity are several orders of magnitude larger; this is partially an artefact of the high sampling rate (300 Hz) used in data acquisition.

Phase coupling within and across legs

We are also interested in the difference between computed phases of joints within a leg and across legs, shown in panels A and B of Figure 7.3, respectively. For phase coupling within a leg, shown in panel A, we choose a representative joint for the leg (denoted "target" on the image), and observe the coupling between this joint and other joints on the leg. Peaks in coupling indicate synchronization: a single peak at zero on the horizontal axis indicates that the two joint phases are coupled to match; a single peak elsewhere indicates that the two joint phases are coupled with some offset. A lack of peaks indicates that the two joints are very weakly or not coupled. We see that in the real data, joints exhibit a mixture of strong and weak coupling, which is largely replicated by the model.

For phase coupling across legs, shown in panel B, we represent each leg with the phase of its representative joint, and compare the phases across legs. Once again, peaks indicate strong coupling, while a lack of peaks indicates a lack of correlation. Compared to intra-leg phase coupling, inter-leg phase coupling appears to have smoother, more well-defined peaks.

7.4   Dynamic Perturbations and Motor Delay

We simulate the model with external dynamic perturbations, and assess the realism of the resulting walking. We consider perturbations that mimic the effects of slippery ground. When the tip of the leg reaches a local minimum (i.e. touches the ground), we inject Gaussian noise into two joints: femur rotation and femur-tibia flexion. We show time series and angle vs. phase data in Figure 7.4, and observe that joint kinematics appear different, but still oscillatory during perturbations. After perturbations cease, joint kinematics recover to pre-perturbation patterns. Thus, the model is able to maintain walking-like behavior in the presence of perturbations, and recovers after perturbations end.

The inclusion of delays in the model allows us to explore varying sensor and motor delays and how this affects walking. Without external perturbations, the model is able to produce realistic walking with arbitrary delays; the model can use the trajectory generator to predict into the future and effectively compensate for delay effects. However, the assumption of perfectly predictable, perturbation-free walking is unrealistic; thus, we consider how motor delays affect walking in the presence of slip-like perturbations.

First, we provide a new summary metric of walking, which will be handy to investigate across varying values of delay and perturbation. For a given bout of walking, we compute the *likelihood* of it occurring in nature (i.e. data). The process to compute this is shown in Figure 7.4. First, a principle component analysis is computed on the set of experiment data. A Gaussian kernel density estimator (KDE) is fitted to this data. This KDE takes a bout as input, and gives the log probability density function (LogPDF) as an output. This is a scalar value corresponding to likelihood, where a near-zero value corresponds to normal walking (i.e. high likelihood of occurring in nature), while a large negative value corresponds to very abnormal walking (i.e. low likelihood of occurring in nature).

We consider both during-perturbation and post-perturbation walking for various delay values and perturbation strengths, and visualize the results in Figure 7.5. We see that for a given perturbation strength, the effect is much more pronounced for higher sensor and motor delays. However, even for large perturbations, the model mostly recovers after perturbations end. Additionally, as expected, the "normalness" of walking increases with increased perturbation strength; however, once delay is below a certain threshold, no noticeable difference in gaits is observed between different perturbation strengths. For instance, during perturbations, a delay of less than 30ms results in relatively normal walking (LogPDF greater than -20). After perturbations, a delay of less than 40ms results in recovery to normal walking (LogPDF of around zero).

7.5 Discussion and Interpretation of Results
Layered model produces robust walking and facilitates local control
One key finding of this model is that robust walking in the presence of external perturbations can organically result from a reflex-like controller layer in combination

with a trajectory generation layer that is pre-programmed for perturbation-free conditions (i.e. trained on data from perturbation-free walking). The model maintains robust walking in the presence of perturbations without requiring re-training. The key enabler of robustness here is the reflex-like controller layer, which rejects perturbations while maintaining a walking gait. In nature, many external perturbations are encountered during legged locomotion; this layered model suggests how robust locomotion may be maintained using a reflex layer, requiring minimal adaptation from non-reflex neurons.

The separation of functions between layers in this model also reinforce that locomotion can be produced by mostly local (i.e. per-leg) signals. In this model, the only signals that need to be communicated between legs are signals about per-leg phases; no information about individual joints needs to be included.

Fundamental constraints on motor delay

This model includes motor delays, which are ubiquitous in animal locomotion due to the biophysics of muscles and neurons. Their values are governed by two opposing design principles. Firstly, from an energy perspective, it is more expensive to manufacture and maintain muscles and neurons with low delay [78] — thus, higher delays are preferable. In contrast, from a performance perspective, delays result in performance degradation and slower reactions, which may impede survival. As seen in our simulations, increased delays lead to worse walking — thus, lower delays are preferable. Our work reconciles these two design principles by showing that evolution selects the maximum value of delay that will still preserve performance (i.e. walking) under reasonable perturbations. In particular, our model predicts that the maximum allowable motor delay is from 20-40ms, which coincides with known values for *Drosophila*.

Motor delay necessitates compensatory prediction

This model uses a novel controller formulation from to include sensor and motor delays using Linear Quadratic Gaussian techniques. During model development, we discovered that for the task of tracking a reference trajectory, compensatory prediction is necessary for good model performance in the presence of motor delays. In all simulations, we allow the controller to use a prediction horizon that matches the motor delay. For instance, if the motor delay is 10 timesteps, then the controller has access to the planned trajectory (from the trajectory generator) up to 10 timesteps into the future. A motor delay of 10 timesteps means that the current planned motor action will not take effect until 10 timesteps in the future, so intuitively it makes sense that we should know the planned trajectory at this point in the future. We also experimented with altering the prediction horizon to be less than the muscle delay, with catastrophic consequences — the model mostly produced only noise. Overall, the model suggests that future predictions are crucial in compensating for motor delays, an idea also proposed in [110].

General framework for models of animal locomotion

The general framework of a three-layer model for locomotion is applicable to any multi-limbed organism for any modality (e.g. flying, swimming). The key ingredients required for this model are: (1) a functional inter-limb coordinator, (2) sufficient data to train a trajectory-generator layer, and (3) a controller that adequately tracks the trajectory for some dynamical model of the organism. The dynamical model may be linearized (as is done in this paper) and controlled with a standard controller, or nonlinear techniques such as feedback linearization may be employed. Overall, this framework allows scientists and engineers to build upon existing models of inter-limb coupling and take advantage of various emerging datasets on animal locomotion to create more fully integrated models of locomotion for various organisms.

7.6   Conclusions and Future Work

We have proposed a layered model of *Drosophila* locomotion. The focus of this model is to produce realistic behavior with reasonable physiological considerations, i.e. link-and-joint dynamics, sensor and motor delays. In order to do so, we have made a number of physiological simplifications. Though quantitative insights are made in the study, numerical values may be sensitive to these simplifications. In particular, ground contact interactions, muscle models, and proprioceptor models are not explicitly included in the dynamics, though they are implicitly taken into account by the trajectories learned by the neural network. The incorporation of these additional features, as well as other biomechanical details, are the subject of future work. We anticipate that the inclusion of proprioceptor models will make the system slightly more difficult to control and therefore lower the allowable values of delay; while the inclusion of muscle models and other biomechanical details will make the system slightly easier to control and therefore increase the allowable values of delay. The reasoning for the latter is that biomechanics (for instance, compliance in the tarsus) are typically somewhat optimized for locomotion.

Another key simplification made by this model is the omission of dynamical coupling. Currently, the legs are only coupled neurally, through the phase coordinator — however, in real life, the legs are also dynamically coupled through the body and its weight distribution upon the legs. This is also a feature we plan to incorporate in future work, as we move toward a more biomechanically realistic model. One avenue of planned investigation is integration with a physics-based model [103]. We anticipate that the inclusion of dynamical coupling may require some additional coordination between the legs. Furthermore, the current controller uses a position-based method, while the inclusion of dynamical coupling will necessitate a switch to impedance-based methods.

This model also has potential applications for bio-mimetic robotic locomotion. In the interests of parsimony, the current model includes a fairly basic dynamical model consisting of linearized link and joints; however, we could also replace this dynamical model with that of a hexapod robot, and recompute a controller accordingly. Due to the modular nature of the model, other modules (trajectory generator, inter-leg coordinator) could remain unchanged — and the resulting code would, in theory, generate bio-mimetic gaits for some robot

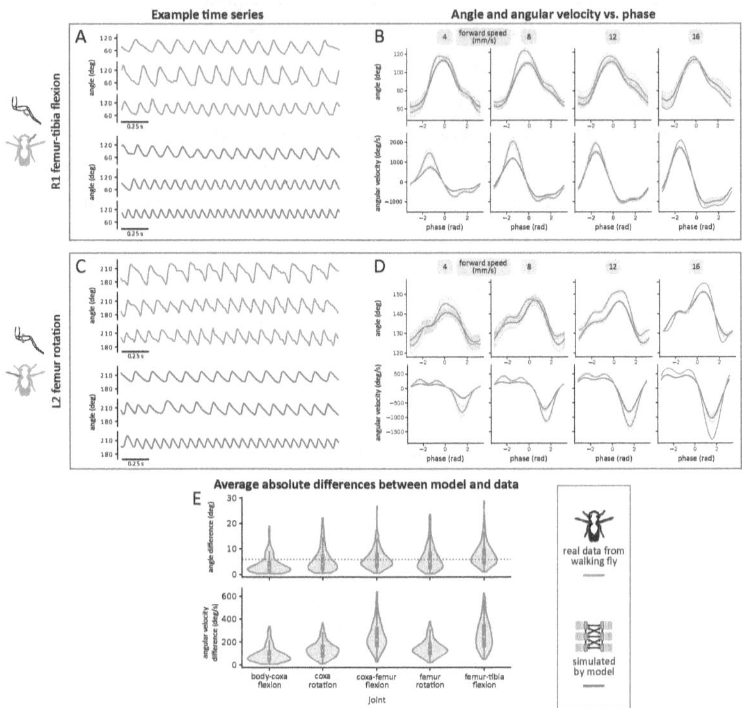

Figure 7.2: Comparison of walking behavior generated by the model (blue) vs. walking behavior recorded from real flies (orange). **(A)** Example time-series of femur-tibia flexion R1 for three different walking speeds. Real data exhibits slightly more variability than model simulations. **(B)** Angle and angular velocity vs. computed per-leg phase of femur-tibia flexion R1 for four different walking speeds. **(C)** Example time-series of femur rotation for leg L2 for three different walking speeds. **(D)** Angle and angular velocity vs. computed per-leg phase for femur rotation for leg L2 for four different walking speeds. **(E)** Average differences between model simulations and data. The dotted line indicates average in-sample differences between different bouts of real walking — we see that the average difference between the model and data is comparable to the in-sample differences in the data itself.

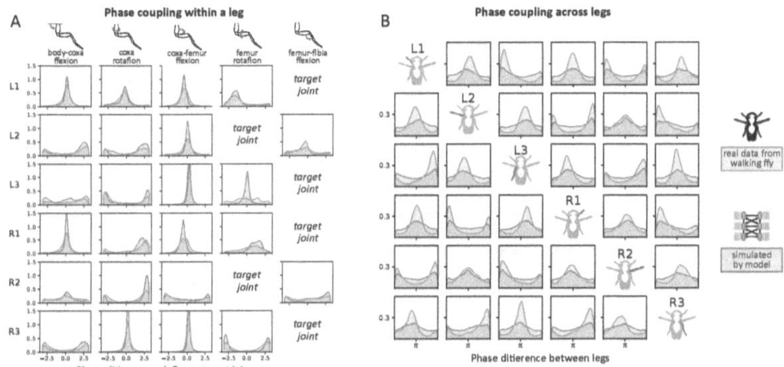

Figure 7.3: Phase coupling within and between legs. **(A)** Phase coupling within each leg. For each leg, we compare phases between a representative joint for the leg (denoted "target" on the image) and other joints on the leg. Peaks in coupling indicate synchronization. The model exhibits slightly stronger coupling than data. **(B)** Phase coupling across legs. We compare phases of representative joints across legs. The model exhibits slightly weaker coupling than data.

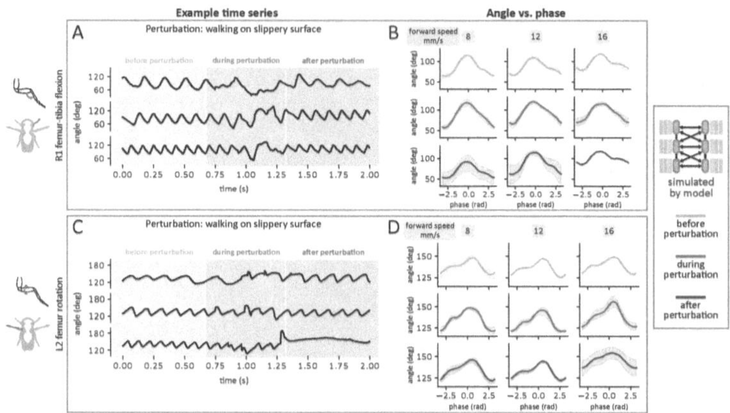

Figure 7.4: Model simulation behavior under perturbations corresponding to walking on a slippery surface. **(A)** Example time-series of femur-tibia flexion on leg R1 for three different walking speeds before, during, and after perturbation. The plot is visibly different during perturbation, but recovers to normal levels after perturbation. **(B)** Angle vs. phase plots for walking before, during, and after perturbation for femur-tibia flexion on leg R1. The plots corresponding to "before" and "after" perturbations greatly resemble one another, suggesting that walking has largely recovered. The plots corresponding to "during perturbations" exhibit a markedly different shape. **(C)** Example time-series of femur rotation for leg L2 for three different walking speeds before, during, and after perturbation. **(D)** Angle vs. phase plots for walking before, during, and after perturbation for femur rotation on leg L2. Similar observations from panel (B) apply.

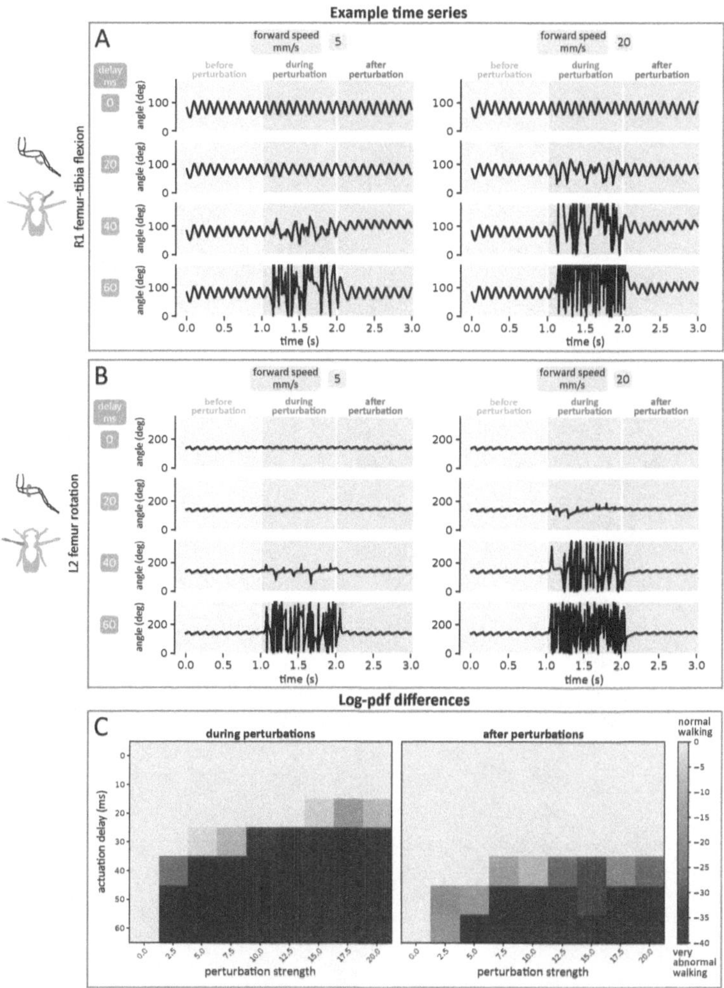

Figure 7.5: Model walking under various perturbations and motor delays. **(A, B)** Example time-series of femur-tibia flexion on leg R1 and femur rotation on leg L2 for various speeds and motor delays, with identical perturbation strength. **(C)** LogPDF differences between perturbed and after-perturbation gaits and perturbation-free walking. During perturbations, a delay of less than 30ms results in relatively normal walking. After perturbations, a delay of less than 40ms results in recovery to normal walking.